Primer on Optimal Control Theory

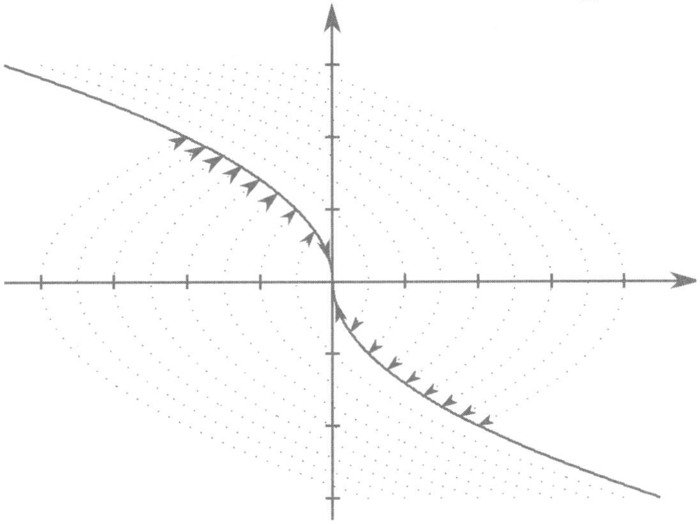

Advances in Design and Control

SIAM's Advances in Design and Control series consists of texts and monographs dealing with all areas of design and control and their applications. Topics of interest include shape optimization, multidisciplinary design, trajectory optimization, feedback, and optimal control. The series focuses on the mathematical and computational aspects of engineering design and control that are usable in a wide variety of scientific and engineering disciplines.

Editor-in-Chief

Ralph C. Smith, North Carolina State University

Editorial Board

Athanasios C. Antoulas, Rice University
Siva Banda, Air Force Research Laboratory
Belinda A. Batten, Oregon State University
John Betts, The Boeing Company (retired)
Stephen L. Campbell, North Carolina State University
Eugene M. Cliff, Virginia Polytechnic Institute and State University
Michel C. Delfour, University of Montreal
Max D. Gunzburger, Florida State University
J. William Helton, University of California, San Diego
Arthur J. Krener, University of California, Davis
Kirsten Morris, University of Waterloo
Richard Murray, California Institute of Technology
Ekkehard Sachs, University of Trier

Series Volumes

Speyer, Jason L., and Jacobson, David H., *Primer on Optimal Control Theory*
Betts, John T., *Practical Methods for Optimal Control and Estimation Using Nonlinear Programming, Second Edition*
Shima, Tal and Rasmussen, Steven, eds., *UAV Cooperative Decision and Control: Challenges and Practical Approaches*
Speyer, Jason L. and Chung, Walter H., *Stochastic Processes, Estimation, and Control*
Krstic, Miroslav and Smyshlyaev, Andrey, *Boundary Control of PDEs: A Course on Backstepping Designs*
Ito, Kazufumi and Kunisch, Karl, *Lagrange Multiplier Approach to Variational Problems and Applications*
Xue, Dingyü, Chen, YangQuan, and Atherton, Derek P., *Linear Feedback Control: Analysis and Design with MATLAB*
Hanson, Floyd B., *Applied Stochastic Processes and Control for Jump-Diffusions: Modeling, Analysis, and Computation*
Michiels, Wim and Niculescu, Silviu-Iulian, *Stability and Stabilization of Time-Delay Systems: An Eigenvalue-Based Approach*
Ioannou, Petros and Fidan, Barış, *Adaptive Control Tutorial*
Bhaya, Amit and Kaszkurewicz, Eugenius, *Control Perspectives on Numerical Algorithms and Matrix Problems*
Robinett III, Rush D., Wilson, David G., Eisler, G. Richard, and Hurtado, John E., *Applied Dynamic Programming for Optimization of Dynamical Systems*
Huang, J., *Nonlinear Output Regulation: Theory and Applications*
Haslinger, J. and Mäkinen, R. A. E., *Introduction to Shape Optimization: Theory, Approximation, and Computation*
Antoulas, Athanasios C., *Approximation of Large-Scale Dynamical Systems*
Gunzburger, Max D., *Perspectives in Flow Control and Optimization*
Delfour, M. C. and Zolésio, J.-P., *Shapes and Geometries: Analysis, Differential Calculus, and Optimization*
Betts, John T., *Practical Methods for Optimal Control Using Nonlinear Programming*
El Ghaoui, Laurent and Niculescu, Silviu-Iulian, eds., *Advances in Linear Matrix Inequality Methods in Control*
Helton, J. William and James, Matthew R., *Extending H^∞ Control to Nonlinear Systems: Control of Nonlinear Systems to Achieve Performance Objectives*

Primer on Optimal Control Theory

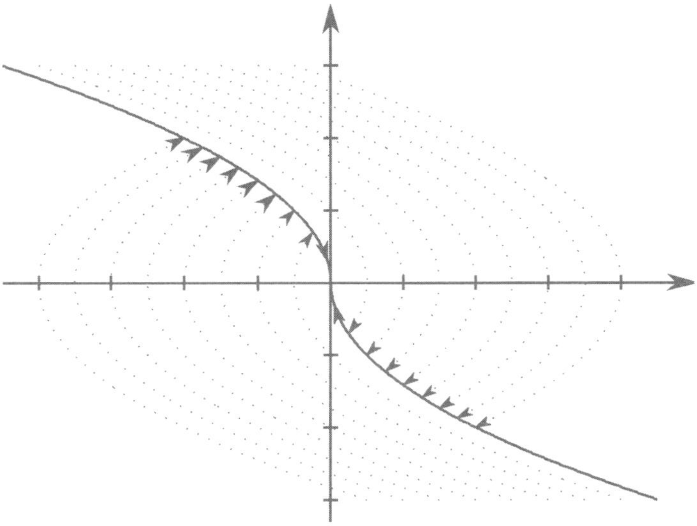

Jason L. Speyer
University of California
Los Angeles, California

David H. Jacobson
PricewaterhouseCoopers LLP
Toronto, Ontario, Canada

siam®

Society for Industrial and Applied Mathematics
Philadelphia

10 9 8 7 6 5 4 3 2 1

Library of Congress Cataloging-in-Publication Data

Speyer, Jason Lee.
Primer on optimal control theory / Jason L. Speyer, David H. Jacobson.
p. cm.
Includes bibliographical references and index.
ISBN 978-0-898716-94-8
1. Control theory. 2. Mathematical optimization. I. Jacobson, David H. II. Title.
QA402.3.S7426 2010
515'.642–dc22
2009047920

To Barbara, a constant source of love and inspiration.
To my children, Gil, Gavriel, Rakhel, and Joseph,
for giving me so much joy and love.

For Celia, Greta, Jonah, Levi, Miles, Thea,
with love from Oupa!

Contents

List of Figures

Preface

This book began when David Jacobson wrote the first draft of Chapters 1, 3, and 4 and Jason Speyer wrote Chapters 2, 5, and 6. Since then the book has constantly evolved by modification of those chapters as we interacted with colleagues and students. We owe much to them for this polished version. The objective of the book is to make optimal control theory accessible to a large class of engineers and scientists who are not mathematicians, although they have a basic mathematical background, but who need to understand and want to appreciate the sophisticated material associated with optimal control theory. Therefore, the material is presented using elementary mathematics, which is sufficient to treat and understand in a rigorous way the issues underlying the limited class of control problems in this text. Furthermore, although many topics that build on this foundation are covered briefly, such as inequality constraints, the singular control problem, and advanced numerical methods, the foundation laid here should be adequate for reading the rich literature on these subjects.

We would like to thank our many students whose input over the years has been incorporated into this final draft. Our colleagues also have been very influential in the approach we have taken. In particular, we have spent many hours discussing the concepts of optimal control theory with Professor David Hull. Special thanks are extended to Professor David Chichka, who contributed some interesting examples and numerical methods, and Professor Moshe Idan, whose careful and critical reading of the manuscript has led to a much-improved final draft. Finally, the first author must express his gratitude to Professor Bryson, a pioneer in the development of the theory, numerical methods, and application of optimal control theory as well as a teacher, mentor, and dear friend.

Chapter 1
Introduction

The operation of many physical processes can be enhanced if more efficient operation can be determined. Such systems as aircraft, chemical processes, and economies have at the disposal of an operator certain controls which can be modulated to enhance some desired property of the system. For example, in commercial aviation, the best fuel usage at cruise is an important consideration in an airline's profitability. Full employment and growth of the gross domestic product are measures of economic system performance; these may be enhanced by proper modulation of such controls as the change in discount rate determined by the Federal Reserve Board or changes in the tax codes devised by Congress.

The essential features of such systems as addressed here are dynamic systems, available controls, measures of system performance, and constraints under which a system must operate. Models of the dynamic system are described by a set of first-order coupled nonlinear differential equations representing the propagation of the state variables as a function of the independent variable, say, time. The state vector may be composed of position, velocity, and acceleration. This motion is influenced by the inclusion of a control vector. For example, the throttle setting and the aerodynamic surfaces influence the motion of the aircraft. The performance criterion which establishes the effectiveness of the control process on the dynamical system can take many forms. For an aircraft, desired performance might be efficient fuel cruise (fuel per range), endurance (fuel per time), or time to a given altitude. The performance criterion is to be optimized subject to the constraints imposed by the system dynamics and other constraints. An important class of constraints are those imposed at the termination of the path. For example, the path of an aircraft may terminate in minimum time at a given altitude and velocity. Furthermore, path constraints that are functions of the controls or the states or are functions of both the state and control vectors may be imposed. Force constraints or maximum-altitude constraints may be imposed for practical implementation.

In this chapter, a simple dynamic example is given to illustrate some of the concepts that are described in later chapters. These concepts as well as the optimization concepts for the following chapters are described using elementary mathematical ideas. The objective is to develop a mathematical structure which can be justified rigorously using elementary concepts. If more complex or sophisticated ideas are required, the reader will be directed to appropriate references. Therefore, the treatment here is not the most general but does cover a large class of optimization problems of practical concern.

1.1 Control Example

A control example establishes the notion of control and how it can be manipulated to satisfy given goals. Consider the forced harmonic oscillator described as

$$\ddot{x} + x = u, \quad x(0), \ \dot{x}(0) \ \text{given}, \tag{1.1}$$

where x is the position. The overdot denotes time differentiation; that is, $\dot{x}$ is dx/dt. This second-order linear differential equation can be rewritten as two first-order differential equations by identifying $x_1 = x$ and $x_2 = \dot{x}$. Then

$$\dot{x}_1 = x_2, \qquad x_1(0) \text{ given}, \tag{1.2}$$

$$\dot{x}_2 = -x_1 + u, \qquad x_2(0) \text{ given}, \tag{1.3}$$

or

$$\begin{bmatrix} \dot{x}_1 \\ \dot{x}_2 \end{bmatrix} = \begin{bmatrix} 0 & 1 \\ -1 & 0 \end{bmatrix} \begin{bmatrix} x_1 \\ x_2 \end{bmatrix} + \begin{bmatrix} 0 \\ 1 \end{bmatrix} u. \tag{1.4}$$

Suppose it is desirable to find a control which drives x_1 and x_2 to the origin from arbitrary initial conditions. Since system (1.4) is controllable (general comments on this issue can be found in [8]), there are many ways that this system can be driven to the origin. For example, suppose the control is proportional to the velocity such as $u = -Kx_2$, $K > 0$, is a constant. Then, asymptotically the position and velocity converge to zero as $t \to \infty$.

Note that the system converges for any positive value of K. It might logically be asked if there is a *best* value of K. This in turn requires some definition for "best." There is a large number of possible criteria. Some common objectives are to minimize the time needed to reach the desired state or to minimize the effort it takes. A criterion that allows the engineer to balance the amount of error against the effort expended is often useful. One particular formulation of this trade-off is the *quadratic performance index*, specialized here to

$$J_1 = \lim_{t_f \to \infty} \int_0^{t_f} (a_1 x_1^2 + a_2 x_2^2 + u^2) dt, \tag{1.5}$$

where $a_1 > 0$ and $a_2 > 0$, and $u = -Kx_2$ is substituted into the performance criterion. The constant parameter K is to be determined such that the cost criterion is minimized subject to the functional form of Equation (1.4).

We will not solve this problem here. In Chapter 2, the parameter minimization problem is introduced to develop some of the basic concepts that are used in the solution. However, a point to note is that the control u does not have to be chosen a priori, but the best functional form will be produced by the optimization process. That is, the process will (usually) produce a control that is expressed as a function of the state of the system rather than an explicit function of time. This is especially true for the quadratic performance index subject to a linear dynamical system (see Chapters 5 and 6).

Other performance measures are of interest. For example, minimum time has been mentioned for where the desired final state was the origin. For this problem to make sense, the control must be limited in some way; otherwise, infinite effort would be expended and the origin reached in zero time. In the quadratic performance index in (1.5), the limitation came from penalizing the use of control (the term u^2 inside the integral). Another possibility is to explicitly bound the control. This could represent some physical limit, such as a maximum throttle setting or limits to steering.

Here, for illustration, the control variable is bounded as

$$|u| \leq 1. \tag{1.6}$$

In later chapters it is shown that the best solution often lies on its bounds. To produce some notion of the motion of the state variables (x_1, x_2) over time, note that Equations (1.2) and (1.3) can be combined by eliminating time as

$$\frac{dx_1/dt}{dx_2/dt} = \frac{x_2}{(-x_1 + u)} \quad \Rightarrow \quad (-x_1 + u)dx_1 = x_2 dx_2. \tag{1.7}$$

Assuming u is a constant, both sides can be integrated to get

$$(x_1 - u)^2 + x_2^2 = R^2, \tag{1.8}$$

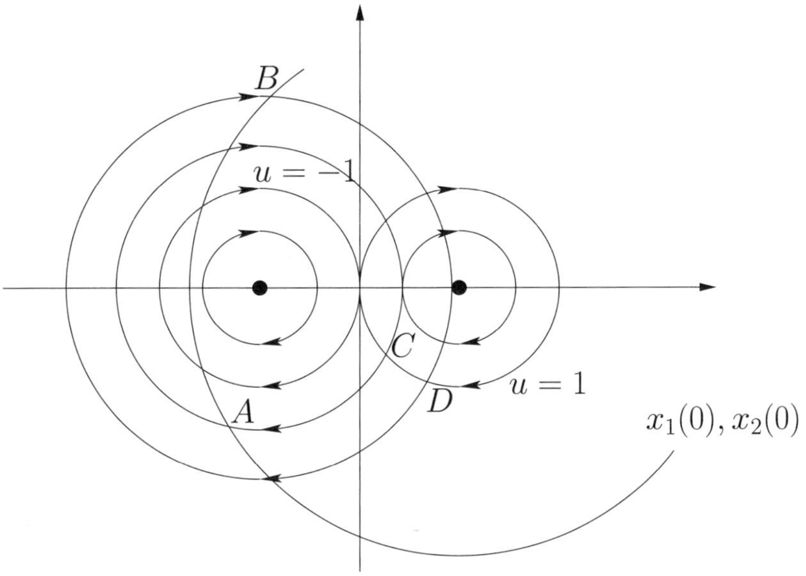

Figure 1.1: Control-constrained optimization example.

which translates to a series of concentric circles for any specific value of the control. For $u = 1$ and $u = -1$, the series of concentric circles are as shown in Figure 1.1 There are many possible paths that drive the initial states $(x_1(0), x_2(0))$ to the origin. Starting with $u = 1$ at some arbitrary $(x_1(0), x_2(0))$, the path proceeds to point A or B. From A or B the control changes, $u = -1$ until point C or D is intercepted. From these points using $u = +1$, the origin is obtained. Neither of these paths starting from the initial conditions is a minimum time path, although starting from point B, the resulting paths are minimum time. The methodology for determining the optimal time paths is given in Chapter 4.

1.2 General Optimal Control Problem

The general form of the optimal control problems we consider begins with a first-order, likely nonlinear, dynamical system of equations as

$$\dot{x} = f(x, u, t), \quad x(t_0) = x_0, \tag{1.9}$$

where $x \in \mathbb{R}^n$, $u \in \mathbb{R}^m$, $f : \mathbb{R}^n \times \mathbb{R}^m \times \mathbb{R}^1 \to \mathbb{R}^n$. Recall that $(\dot{\ })$ denotes $d(\ )/dt$. Denote $x(t)$ as x, $u(t)$ as u, and the functions of x and u as $x(\cdot)$ and $u(\cdot)$. In the example of Section 1.1, the system is given by Equation (1.4).

The performance of the dynamical system is to be modulated to minimize some performance index, which we assume to be of the form

$$J = \phi(x(t_f), t_f) + \int_{t_0}^{t_f} L(x, u, t)dt, \tag{1.10}$$

where $\phi : \mathbb{R}^n \times \mathbb{R}^1 \to \mathbb{R}^1$ and $L : \mathbb{R}^n \times \mathbb{R}^m \times \mathbb{R}^1 \to \mathbb{R}^1$. The terms in the performance index are often driven by considerations of energy use and time constraints. For example, the performance index might be as simple as minimizing the final time (set $\phi(t_f) = t_f$ and $L(\cdot, \cdot, \cdot) \equiv 0$ in (1.10)). It may also attempt to minimize the amount of energy expended in achieving the desired goal or to limit the control effort expended, or any combination of these and many other considerations.

In the formulation of the problem, we limit the class of control functions $\mathcal{U}$ to the class of bounded piecewise continuous functions. The solution is to be such that the functional J takes on its minimum for some $u(\cdot) \in \mathcal{U}$ subject to the differential equations (1.9). There may also be several other constraints.

One very common form of constraint, which we treat at length, is on the terminal state of the system:

$$\psi(x(t_f), t_f) = 0, \tag{1.11}$$

where $\psi : \mathbb{R}^n \times \mathbb{R}^1 \to \mathbb{R}^p$. This reflects a common requirement in engineering problems, that of achieving some specified final condition exactly.

The motion of the system and the amount of control available may also be subject to hard limits. These *bounds* may be written as

$$S(x(t), t) \leq 0, \tag{1.12}$$

where $S : \mathbb{R}^n \times \mathbb{R}^1 \to \mathbb{R}^1$ for a bound on the state only, or more generally for a mixed state and control space bound

$$g(x(t), u(t), t) \leq 0, \tag{1.13}$$

where $g : \mathbb{R}^n \times \mathbb{R}^m \times \mathbb{R}^1 \to \mathbb{R}^1$. These bounds represent physical or other limitations on the system. For an aircraft, for instance, the altitude must always be greater than that of the landscape, and the control available is limited by the physical capabilities of the engines and control surfaces.

Many important classes of problems have been left out of our presentation. For example, the state variable inequality constraint given in (1.12) is beyond the scope of this book.

1.3 Purpose and General Outline

This book aims to provide a treatment of control theory using mathematics at the level of the practicing engineer and scientist. The general problem cannot be treated in complete detail using essentially elementary mathematics. However, important special cases of the general problems can be treated in complete detail using elementary mathematics. These special cases are sufficiently broad to solve many interesting and important problems. Furthermore, these special cases suggest solutions to the more general problem. Therefore, complete solutions to the general problem are stated and used. The theoretical gap between the solution to the special cases and the solution to the general problem is discussed, and additional references are given for completeness.

To introduce important concepts, mathematical style, and notation, in Chapter 2 the parameter minimization problem is formulated and conditions for local optimality are determined. By local optimality we mean that optimality can be verified about a small neighborhood of the optimal point. First, the notions of first- and second-order local necessary conditions for unconstrained parameter minimization problems are derived. The first-order necessary conditions are generalized in Chapter 3 to the minimization of a general performance criterion with nonlinear dynamic systems constraints. Next, the notion of first- and second-order local necessary conditions for parameter minimization problems is extended to include algebraic constraints. The first-order necessary conditions are generalized in Chapter 4 to the minimization of a general performance criterion with nonlinear dynamic systems constraints *and* terminal equality constraints. Second-order local necessary conditions for the minimization of general performance criterion with nonlinear dynamic systems constraints for both unconstrained and terminal equality constrained problems are given in Chapter 5.

In Chapters 3 and 4, local and global conditions for optimality are given for what are called weak and strong control variations. "Weak control variation" means that at any point, the variation away from the optimal control is very small; however, this small variation may be everywhere along the path. This gives rise to the classical local necessary conditions of Euler and Lagrange. "Strong control variation" means that the variation is zero over most of the path, but along a very short section it may be arbitrarily large. This leads to the classical Weierstrass local conditions and its more modern generalization called the Pontryagin Maximum Principle. The local optimality conditions are useful in constructing numerical algorithms for determining the optimal path. Less useful numerically, but sometimes very helpful theoretically, are the global sufficiency conditions. These necessary conditions require the solution to a partial differential equation known as the Hamilton–Jacobi–Bellman equation.

In Chapter 5 the second variation for weak control variations produces local necessary and sufficient conditions for optimality. These conditions are determined by solving what is called the accessory problem in the calculus of variations, which is essentially minimizing a quadratic cost criterion subject to linear differential equations, i.e., the linear quadratic problem. The linear quadratic problem also arises directly and naturally in many applications and is the basis of much control synthesis work. In Chapter 6 the linear quadratic problem of Chapter 5 is generalized to a two-sided optimization problem producing a zero-sum differential game. The solutions to both the linear quadratic problem and the zero-sum differential game problem produce linear feedback control laws, known in the robust control literature as the H_2 and H_∞ controllers.

Background material is included in the appendix. The reader is assumed to be familiar with differential equations and standard vector-matrix algebra.

Chapter 2
Finite-Dimensional Optimization

A popular approach to the numerical solution of functional minimization problems, where a piecewise continuous control function is sought, is to convert them to an approximate parameter minimization problem. This motivation for the study of parameter minimization is shown more fully in Section 2.1. However, many of the ideas developed to characterize the parameter optimal solution extend to the functional optimization problem but can be treated from a more transparent viewpoint in this setting. These include the first-order and second-order necessary and sufficient conditions for optimality for both unconstrained and constrained minimization problems.

2.1 Motivation for Considering Parameter Minimization for Functional Optimization

Following the motivation given in Chapter 1, we consider the functional optimization problem of minimizing with respect to $u(\cdot) \in \mathcal{U}$,[1]

$$J(u, x_0) = \phi(x(t_f), t_f) + \int_{t_0}^{t_f} L(x(t), u(t), t)\,dt \tag{2.1}$$

[1]$\mathcal{U}$ represents the class of bounded piecewise continuous functions.

subject to

$$\dot{x}(t) = f(x(t), u(t), t), \quad x_0 \text{ given.} \tag{2.2}$$

This functional optimization problem can be converted to a parameter optimization or function optimization problem by assuming that the control is piecewise linear as

$$u(t) = \hat{u}(u_p, t) = u_i(t_i) + \frac{(t - t_i)}{t_{i+1} - t_i}(u_{i+1} - u_i), \quad t_i \le t \le t_{i+1}, \tag{2.3}$$

where $i = 0, \ldots, N-1, t_f = t_N$, and we define the parameter vector as

$$u_p = \{u_i, \quad i = 0, \ldots, N-1\}. \tag{2.4}$$

The optimization problem is then as follows. Find the control $\hat{u}(\cdot) \in \mathcal{U}$ that minimizes

$$J(\hat{u}, x_0) = \phi(x(t_f), t_f) + \int_{t_0}^{t_f} L(x, \hat{u}(u_p, t), t)\, dt \tag{2.5}$$

subject to

$$\dot{x} = f(x(t), \hat{u}(u_p, t), t), \quad x(0) = x_0 \text{ given.} \tag{2.6}$$

Thus, the functional minimization problem is transformed into a parameter minimization problem to be solved over the time interval $[t_0, t_f]$. Since the solution to (2.6) is the state as a function of u_p, i.e., $x(t) = \hat{x}(u_p, t)$, then the cost criterion is

$$J(\hat{u}(u_p), x_0) = \phi(\hat{x}(u_p, t_f)) + \int_{t_0}^{t_f} L(\hat{x}(u_p, t), \hat{u}(u_p, t), t)\, dt, \tag{2.7}$$

The parameter minimization problem is to minimize $J(\hat{u}(u_p), x_0)$ with respect to u_p. Because we have made assumptions about the form of the control function, this will produce a result that is suboptimal. However, when care is taken, the result will be close to optimal.

Example 2.1.1 *As a simple example, consider a variant of the brachistochrone problem, first proposed by John Bernoulli in 1696. As shown in Figure* 2.1*, a bead is sliding on a wire from an initial point O to some point on the wall at a known $r = r_f$. The wire is frictionless. The problem is to find the shape of the wire such that the bead arrives at the wall in minimum time.*

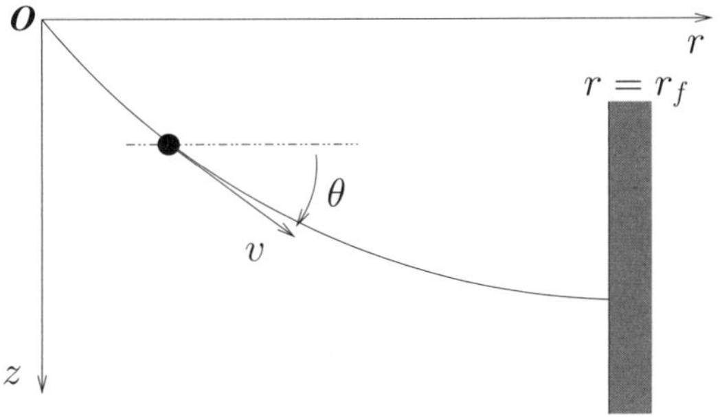

Figure 2.1: A brachistochrone problem.

In this problem, the control function is $\theta(t)$, and the system equations are

$$\begin{aligned} \dot{z} &= v \sin\theta, \qquad z(0) = 0,\ z(t_f)\ \text{free}, \\ \dot{r} &= v \cos\theta, \qquad r(0) = 0,\ r(t_f) = 1, \\ \dot{v} &= g \sin\theta, \qquad v(0) = 0,\ v(t_f)\ \text{free}, \end{aligned}$$

where g is the constant acceleration due to gravity, and the initial point O is taken to be the origin. The performance index to be minimized is simply

$$J(\hat{\theta}, O) = t_f.$$

The control can be parameterized in this case as a function of r more easily than as a function of time, as the final time is not known. To make the example more concrete, let $r_f = 1$ and assume a simple approximation by dividing the interval into halves, with the parameters being the slopes at the beginning, midpoint, and end,

$$u_p = \{u_0,\ u_1,\ u_2\} = \{\theta(0),\ \theta(0.5),\ \theta(1)\},$$

so that

$$\hat{\theta}(r) = \begin{cases} u_0 + \frac{r}{0.5}(u_1 - u_0), & 0 \le r \le 0.5, \\ u_1 + \frac{r-0.5}{0.5}(u_1 - u_2), & 0.5 < r \le 1. \end{cases}$$

The problem is now converted to minimization of the final time over these three independent variables.

In the next sections, we develop the theory of parameter optimization.

2.2 Unconstrained Minimization

Consider that the cost criterion is a scalar function $\phi(\cdot)$ of a single variable x for $x \in [x_a, x_b]$.

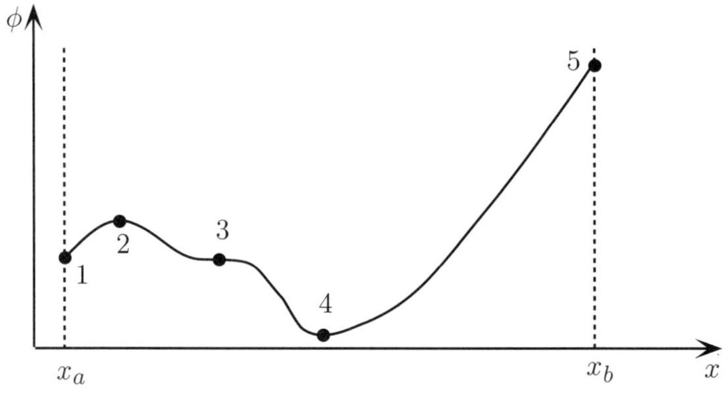

Figure 2.2: Definition of extremal points.

The interior extremal points for this function are, as shown in Figure 2.2, 2, a relative maximum; 3, an inflection (saddle) point; and 4, a relative minimum (absolute for $x \in [x_a, x_b]$). The boundary point extrema are 1, a relative minimum, and 5, a and relative maximum (absolute for $x \in [x_a, x_b]$).

Remark 2.2.1 *We consider first only interior extremal points.*

Assumption 2.2.1 *Assume that $\phi(x)$ is continuously differentiable everywhere in $[x_a, x_b]$.*

The assumption avoids functions as shown in Figure 2.3.

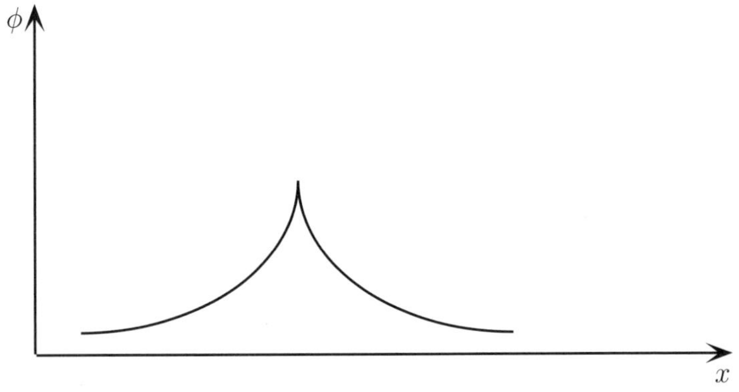

Figure 2.3: Function with a discontinuous derivative.

2.2.1 Scalar Case

To focus on the essential notions of determining both local first- and second-order necessary and sufficiency conditions, a scalar problem is used. These ideas are applied throughout the book.

First-Order Necessary Conditions

The following theorem and its proof sets the style and notation for the analysis that is used in more complex problems.

Theorem 2.2.1 *Let $\Omega = (x_a, x_b)$. Let the cost criterion $\phi : \mathbb{R} \to \mathbb{R}$ be a differentiable function. Let x^o be an optimal solution of the optimization problem*

$$\min_x \phi(x) \quad \textit{subject to} \quad x \in \Omega. \tag{2.8}$$

Then it is necessary that

$$\frac{\partial \phi}{\partial x}(x^o) = 0. \tag{2.9}$$

Remark 2.2.2 *This is a first-order necessary condition for stationarity (a local (relative) minimum, a maximum, or a saddle (inflection) point; see Figure* 2.2*).*

Proof: Since $x^o \in \Omega$ and Ω is an *open interval*, there exists an $\epsilon > 0$ such that $x \in \Omega$ whenever $|x - x^o| < \epsilon$. This implies that for all $x^o + \gamma\alpha \in \Omega$, where for any $\alpha \in \mathbb{R}$ and $0 \le \gamma \le \eta$ (η is determined by α), we have

$$\phi(x^o) \le \phi(x^o + \gamma\alpha). \tag{2.10}$$

Since ϕ is differentiable, by Taylor's Theorem (see Appendix A.1.2)

$$\phi(x^o + \gamma\alpha) = \phi(x^o) + \gamma\frac{\partial\phi}{\partial x}(x^o)\alpha + \mathcal{O}(\gamma), \tag{2.11}$$

where $\mathcal{O}(\gamma)$ denotes terms of order greater than γ such that

$$\frac{\mathcal{O}(\gamma)}{\gamma} \to 0 \quad \text{as} \quad \gamma \to 0. \tag{2.12}$$

Substitution of (2.11) into (2.10) yields

$$0 \le \gamma\frac{\partial\phi}{\partial x}(x^o)\alpha + \mathcal{O}(\gamma). \tag{2.13}$$

Dividing this by $\gamma > 0$ gives

$$0 \le \frac{\partial\phi}{\partial x}(x^o)\alpha + \frac{\mathcal{O}(\gamma)}{\gamma}. \tag{2.14}$$

Let $\gamma \to 0$ to yield

$$0 \le \frac{\partial\phi}{\partial x}(x^o)\alpha. \tag{2.15}$$

Since the inequality must hold for all α, in particular for both positive and negative values of α, then this implies

$$\frac{\partial\phi}{\partial x}(x^o) = 0. \tag{2.16}$$

∎

Second Variation

Suppose ϕ is twice differentiable and let $x^o \in \Omega$ be an optimal or even a locally optimal solution. Then $\partial\phi/\partial x(x^o) \triangleq \phi_x(x^o) = 0$, and by Taylor's Theorem

$$\phi(x^o + \gamma\alpha) = \phi(x^o) + \frac{1}{2}\gamma^2\phi_{xx}(x^o)\alpha^2 + \mathcal{O}(\gamma^2), \tag{2.17}$$

where

$$\frac{\mathcal{O}(\gamma^2)}{\gamma^2} \to 0 \quad \text{as} \quad \gamma \to 0. \tag{2.18}$$

For γ sufficiently small,

$$\phi(x^o) \le \phi(x^o + \gamma\alpha) = \phi(x^o) + \frac{1}{2}\gamma^2\phi_{xx}(x^o)\alpha^2 + \mathcal{O}(\gamma^2), \tag{2.19}$$

$$\Rightarrow \quad 0 \le \frac{1}{2}\phi_{xx}(x^o)\alpha^2 + \frac{\mathcal{O}(\gamma^2)}{\gamma^2} \tag{2.20}$$

after dividing by $\gamma^2 > 0$. For $\gamma^2 \to 0$, this yields

$$\frac{1}{2}\phi_{xx}(x^o)\alpha^2 \ge 0 \tag{2.21}$$

for all α. This means that $\phi_{xx}(x^o)$ is nonnegative (see Appendix A.2.5 for a discussion on quadratic forms and definite matrices) and is another *necessary condition.* Equation (2.21) is known as a second-order necessary condition or a convexity condition.

Sufficient Condition for a Local Minimum

Suppose that $x^o \in \Omega$, $\phi_x(x^o) = 0$, and $\phi_{xx}(x^o) > 0$ (strictly positive). Then from

$$\phi(x^o) < \phi(x^o + \gamma\alpha) \quad \text{for all } x^o + \gamma\alpha \in \Omega, \tag{2.22}$$

we can conclude that x^o is a local minimum (see Figure 2.2).

Remark 2.2.3 *If the second variation dominates all other terms in the Taylor series (2.19), then it is called strongly positive, and $\phi_x(x^o) = 0$ and $\phi_{xx}(x^o) > 0$ are* *necessary and sufficient conditions for a local minimum. This concept becomes nontrivial in functional optimization, discussed in Chapter* 5.

Higher-Order Variations

For $\alpha \in \mathbb{R}$ and $0 \le \gamma \le \eta$ denote the change in ϕ as

$$\Delta\phi \stackrel{\triangle}{=} \phi(x^o + \gamma\alpha) - \phi(x^o). \tag{2.23}$$

Expanding this into a Taylor series gives

$$\Delta\phi = \delta\phi + \frac{1}{2}\delta^2\phi + \frac{1}{3!}\delta^3\phi + \frac{1}{4!}\delta^4\phi + \cdots, \tag{2.24}$$

where

$$\delta\phi = \phi_x(x^o)\gamma\alpha, \quad \delta^2\phi = \phi_{xx}(x^o)(\gamma\alpha)^2, \text{ etc.} \tag{2.25}$$

Suppose that $\phi_x(x^o) = 0$ and also $\phi_{xx}(x^o) = 0$. If $\phi_{xxx}(x^o) \neq 0$, the extremal is a saddle. If $\phi_{xxx}(x^o) = 0$ and $\phi_{xxxx}(x^o) > 0$, the extremal is a local minimum. These conditions can be seen, respectively, in the examples $\phi(x) = x^3$ and $\phi(x) = x^4$.

Note that the conditions for a maximum can be obtained from those for a minimum by replacement of ϕ by $-\phi$. Hence $\phi_x(x^o) = 0$, $\phi_{xx}(x^o) \le 0$, are necessary conditions for a local maximum, and $\phi_x(x^o) = 0$, $\phi_{xx}(x^o) < 0$, are sufficient conditions for a local maximum.

2.2.2 Numerical Approaches to One-Dimensional Minimization

In this section we present two common numerical methods for finding the point at which a function is minimized. This will clarify what has just been presented. We make tacit assumptions that the functions involved are well behaved and satisfy continuity and smoothness conditions. For more complete descriptions of numerical optimization, see such specialized texts as [23] and [36].

Golden Section Searches

Suppose that it is known that a minimum of the function $\phi(x)$ exists on the interval (a, b). The only way to be certain that an interval (a, b) contains a minimum is to have some $\bar{x} \in (a, b)$ such that $\phi(\bar{x}) < \phi(a)$ and $\phi(\bar{x}) < \phi(b)$. Assuming that there is only one minimum in the interval, the first step in finding its precise location is to find whether the minimizer is in one of the subintervals $(a, \bar{x}]$ or $[\bar{x}, b)$. (The subintervals are partly closed because it is possible that $\bar{x}$ is the minimizer.)

To find out, we apply the same criterion to one of the subintervals. That is, we choose a test point $x_t \in (a, b), x_t \neq \bar{x}$, and evaluate the function at that point. Suppose that $x_t < \bar{x}$. We can then check to see if $\phi(x_t) < \phi(\bar{x})$. If it is, we know that the minimum lies in the interval $(a, \bar{x})$. If $\phi(\bar{x}) < \phi(x_t)$, then the minimum must lie in the interval (x_t, b). Note that due to our strong assumption about a single minimum, $\phi(\bar{x}) = \phi(x_t)$ implies that the minimum is in the interval $(x_t, \bar{x})$.

What is special about the golden section search is the way in which the test points are chosen. The golden ratio has the value

$$\mathcal{G} = \frac{\sqrt{5} - 1}{2} \approx 0.61803\ldots.$$

Given the points a and b bracketing a minimum, we choose two additional points x_1 and x_2 as

$$x_1 = b - \mathcal{G}(b - a), \quad x_2 = a + \mathcal{G}(b - a),$$

which gives us four points in the order a, x_1, x_2, b. Now suppose that $\phi(x_1) < \phi(x_2)$. Then we know that the location of the minimum is between a and x_2. Conversely, if $\phi(x_2) < \phi(x_1)$, the minimum lies between x_1 and b. In either case, we are left with three points, and the interior of these points is already in the right position to be used in the next iteration. In the first case, for example, the new interval is (a, x_2), and the point x_1 satisfies the relationship $x_1 = a + \mathcal{G}(x_2 - a)$.

This leads to the following algorithm:

Given the points a, x_1, x_2, and b and the corresponding values of the function, then

1. If $\phi(x_1) \leq \phi(x_2)$, then
   (a) Set $b = x_2$, and $\phi(b) = \phi(x_2)$.
   (b) Set $x_2 = x_1$, and $\phi(x_2) = \phi(x_1)$.
   (c) Set $x_1 = b - \mathcal{G}(b - a)$, and compute $\phi(x_1)$. (*Note*: Use the value of b *after* updating as in 1(a).)
2. Else
   (a) Set $a = x_1$, and $\phi(a) = \phi(x_1)$.
   (b) Set $x_1 = x_2$, and $\phi(x_1) = \phi(x_2)$.
   (c) Set $x_2 = a + \mathcal{G}(b - a)$, and compute $\phi(x_2)$. (Note: Use the value of a *after* updating as in 1(a).)
3. If the length of the interval is sufficiently small, then
   (a) If $\phi(x_1) \leq \phi(x_2)$, return x_1 as the minimizer.
   (b) Else return x_2 as the minimizer.
4. Else go to 1.

Note: The assumption that the function is well behaved impies that at least one of $\phi(x_1) < \phi(a)$ or $\phi(x_2) < \phi(b)$ is true. Furthermore, "well behaved" implies that the second derivative $\phi_{xx}(\bar{x}) > 0$ and that $\phi_x = 0$ only at $\bar{x}$ on the interval.

Newton Iteration

The golden section search is simple and reliable. However, it requires knowledge of an interval containing the minimum. It also converges linearly; that is, the size of the interval containing the minimum is reduced by the same ratio (in this case $\mathcal{G}$) at each step.

Consider instead the point $\bar{x}$ and assume that the function can be well approximated by the first few terms of the Taylor expansion about that point. That is,

$$\phi(x) = \phi(\bar{x} + h) \approx \phi(\bar{x}) + \phi_x(\bar{x})h + \frac{\phi_{xx}(\bar{x})}{2}h^2. \tag{2.26}$$

Minimizing this expression over h gives

$$h = -\frac{\phi_x(\bar{x})}{\phi_{xx}(\bar{x})}.$$

The method proceeds iteratively as

$$x^{i+1} = x^i - \frac{\phi_x(x^i)}{\phi_{xx}(x^i)}.$$

It can be shown that near the minimum this method converges quadratically. That is, $|x^{i+1} - x^o| \sim |x^i - x^o|^2$. However, if the assumption (2.26) does not hold, the method will diverge quickly.

2.2.3 Functions of n Independent Variables: First-Order Conditions

In this section the cost criterion $\phi(\cdot)$ to be minimized is a function of an n-vector x. In order to characterize the length of the vector, the notion of a norm is introduced. (See Appendix A.2.9 for a more complete description.) For example, define the Euclidean norm as $\|x\| = (x^T x)^{\frac{1}{2}}$.

Theorem 2.2.2 *Suppose $x \in \mathbb{R}^n$ where $x = [x_1, \ldots, x_n]^T$. Let $\phi(x) : \mathbb{R}^n \to \mathbb{R}$ and be differentiable. Let Ω be an open subset of $\mathbb{R}^n$. Let x^o be an optimal solution to the problem*[2]

$$\min_x \phi(x) \quad \textit{subject to} \quad x \in \Omega. \tag{2.27}$$

Then

$$\left.\frac{\partial \phi}{\partial x}\right|_{x=x^o} = \phi_x(x^o) = 0, \tag{2.28}$$

where $\phi_x(x^o) = [\phi_{x_1}(x^o), \ldots, \phi_{x_n}(x^o)]$ is a row vector.

Proof: Since $x^o \in \Omega$ is an open subset of $\mathbb{R}^n$, then there exists an $\epsilon > 0$ such that $x \in \Omega$ whenever x belongs to an n-dimensional ball $\|x - x^o\| < \epsilon$ (or an n-dimensional box $|x_i - x_i^o| < \epsilon_i, i = 1, \ldots, N$). Therefore, for every vector $\alpha \in \mathbb{R}^n$ there is a $\gamma > 0$ (γ depends upon α) such that

$$(x^o + \gamma\alpha) \in \Omega \quad \text{whenever} \quad 0 \le \gamma \le \eta, \tag{2.29}$$

where η is related to $\|\alpha\|$. Since x^o is optimal, we must then have

$$\phi(x^o) \le \phi(x^o + \gamma\alpha) \quad \text{whenever} \quad 0 \le \gamma \le \eta. \tag{2.30}$$

Since ϕ is once continuously differentiable, by Taylor's Theorem (Equation (A.48)),

$$\phi(x^o + \gamma\alpha) = \phi(x^o) + \phi_x(x^o)\gamma\alpha + \mathcal{O}(\gamma), \tag{2.31}$$

where $\mathcal{O}(\gamma)$ is the remainder term, and $\frac{\mathcal{O}(\gamma)}{\gamma} \to 0$ as $\gamma \to 0$. Substituting (2.31) into the inequality (2.30) yields

$$0 \le \phi_x(x^o)\gamma\alpha + \mathcal{O}(\gamma). \tag{2.32}$$

[2]This implies that $x^o \in \Omega$.

Dividing this by γ and letting $\gamma \to 0$ gives

$$0 \le \phi_x(x^o)\alpha. \tag{2.33}$$

Since the *inequality must hold for all* $\alpha \in \mathbb{R}^n$, we have

$$\phi_x(x^o) = 0. \tag{2.34}$$

∎

Remark 2.2.4 *Note that* $\phi_x(x^o) = 0$ *gives n nonlinear equations with n unknowns,*

$$\phi_{x_1}(x^o) = 0, \ldots, \phi_{x_n}(x^o) = 0. \tag{2.35}$$

This can be solved for x^o*, but it could be a difficult numerical procedure.*

Remark 2.2.5 *Sometimes, instead of* $\gamma\alpha$*, the variation can be written as* $x - x^o = \delta x = \gamma\alpha$*, but instead of dividing by* γ*, we can divide by* $\|\delta x\|$.

2.2.4 Functions of n Independent Variables: Second-Order Conditions

Suppose ϕ *is twice differentiable.* Let $x^o \in \Omega$ be locally minimum. Then $\phi_x(x^o) = 0$ and by Taylor's expansion (see Appendix A.2.10)

$$\phi(x^o + \gamma\alpha) = \phi(x^o) + \frac{1}{2}\gamma^2\alpha^T\phi_{xx}(x^o)\alpha + \mathcal{O}(\gamma^2), \tag{2.36}$$

where $\frac{\mathcal{O}(\gamma^2)}{\gamma^2} \to 0$ as $\gamma \to 0$. Note that $\phi_{xx} \stackrel{\triangle}{=} (\phi_x^T)_x$ is a symmetric matrix

$$\phi_{xx} = \begin{bmatrix} \phi_{x_1x_1} & \cdots & \phi_{x_1x_n} \\ \vdots & & \\ \phi_{x_nx_1} & \cdots & \phi_{x_nx_n} \end{bmatrix}. \tag{2.37}$$

For $\gamma > 0$ sufficiently small,

$$\phi(x^o) \le \phi(x^o + \gamma\alpha) = \phi(x^o) + \frac{1}{2}\gamma^2\alpha^T\phi_{xx}(x^o)\alpha + \mathcal{O}(\gamma^2) \tag{2.38}$$

$$\Rightarrow \quad 0 \le \frac{1}{2}\gamma^2\alpha^T\phi_{xx}(x^o)\alpha + \mathcal{O}(\gamma^2). \tag{2.39}$$

Dividing through by γ^2 and letting $\gamma \to 0$ gives

$$\frac{1}{2}\alpha^T\phi_{xx}(x^o)\alpha \ge 0. \tag{2.40}$$

As shown in Appendix A.2.5, this means that

$$\phi_{xx}(x^o) \ge 0 \tag{2.41}$$

(nonnegative definite). This is a *necessary* condition for a local minimum. The *sufficient* conditions for a local minimum are

$$\phi_x(x^o) = 0, \quad \phi_{xx}(x^o) > 0 \tag{2.42}$$

(positive definite). These conditions are sufficient because the second variation dominates the Taylor expansion, i.e., if $\phi_{xx}(x^o) > 0$ there always exists a γ such that $\mathcal{O}(\gamma^2)/\gamma^2 \to 0$ as $\gamma \to 0$.

Suppose $\phi_{xx}(x^o)$ is positive definite. Then, (2.40) is satisfied by the strict inequality and the quadratic form has a nice geometric interpretation as an n-dimensional ellipsoid defined by $\alpha^T\phi_{xx}(x^o)\alpha = b$, where b is a given positive scalar constant.

Example 2.2.1 *Consider the performance criterion (or performance index)*

$$\phi(x_1, x_2) = \frac{(x_1^2 + x_2^2)}{2}.$$

Application of the first-order necessary conditions gives

$$\phi_{x_1} = 0 \Rightarrow x_1^o = 0, \quad \phi_{x_2} = 0 \Rightarrow x_2^o = 0.$$

Check to see if (x_1^o, x_2^o) *is a minimum. Using the second variation conditions*

$$\phi_{xx} = \begin{bmatrix} \phi_{x_1x_1} & \phi_{x_1x_2} \\ \phi_{x_2x_1} & \phi_{x_2x_2} \end{bmatrix} = \begin{bmatrix} 1 & 0 \\ 0 & 1 \end{bmatrix}$$

is positive definite because the diagonal elements are positive and the determinant of the matrix itself is positive. Alternately, the eigenvalue of ϕ_{xx} *must be positive (see Appendix* A.2.5*).*

Example 2.2.2 *Consider the performance index*

$$\phi(x_1, x_2) = x_1 x_2.$$

Application of the first-order necessary conditions gives

$$\phi_{x_1} = 0 \Rightarrow x_2^o = 0, \quad \phi_{x_2} = 0 \Rightarrow x_1^o = 0.$$

Check to see if (x_1^o, x_2^o) *is a minimum. Using the second variation conditions,*

$$\phi_{xx} = \begin{bmatrix} 0 & 1 \\ 1 & 0 \end{bmatrix} \Rightarrow |\phi_{xx} - \lambda I| = \begin{vmatrix} -\lambda & 1 \\ 1 & -\lambda \end{vmatrix} = \lambda^2 - 1 = 0.$$

Since the eigenvalues $\lambda = 1, -1$ *are mixed in sign, then the matrix* ϕ_{xx} *is called indefinite.*

2.2.5 Numerical Optimization Schemes

Three numerical optimization techniques are described: a first-order method called steepest descent, a second-order method known as the Newton–Raphson method, and a method that is somewhere between these in numerical complexity and rate of converges, denoted here as the accelerated gradient method.

Steepest Descent (or Gradient) Method

A numerical optimization method is presented based on making small perturbations in the cost criterion function about a nominal value of the state vector. Then, small improvements are made iteratively in the value of the cost criterion. These small improvements are constructed by assuming that the functions evaluated with respect to these small perturbations are essentially linear and, thereby, predict the improvement. If the actual change and the predicted change do not match within given tolerances, then the size of the small perturbations is adjusted.

Consider the problem

$$\min_x \phi(x). \tag{2.43}$$

Let x^i be the value of an x vector at the ith iteration. Perturbing x gives

$$\phi(x) - \phi(x^i) = \Delta\phi(x) = \phi_x(x^i)\delta x + \mathcal{O}(||\delta x||), \quad \delta x \triangleq x - x^i. \tag{2.44}$$

Choose $\delta x^i = -\epsilon^i \phi_x^T(x^i)$, such that $x^{i+1} = x^i + \delta x^i$, and

$$\Delta\phi(x^{i+1}) = -\epsilon^i \phi_x(x^i)\phi_x^T(x^i) + \mathcal{O}(\epsilon^i), \tag{2.45}$$

where the value chosen for ϵ^i is sufficiently small so that the assumed linearity remains valid and the cost criterion decreases as shown in (2.45). As the local minimum is approached, the gradient converges as

$$\lim_{i\to\infty} \phi_x(x^i) \to 0. \tag{2.46}$$

For a quadratic function, the steepest descent method converges in an infinite number of steps. This is because the step size, as expressed by its norm $||\delta x^i|| = \epsilon^i ||\phi_x^T(x^i)||$, becomes vanishingly small.

Newton–Raphson Method

Assume that near the minimum the gradient method is converging slowly. To correct this, expand $\phi(x)$ to second order about the iteration value x^i as

$$\Delta\phi(x) = \phi_x(x^i)\delta x + \frac{1}{2}\delta x^T \phi_{xx}(x^i)\delta x + \mathcal{O}(||\delta x||^2), \tag{2.47}$$

where $\delta x = x - x^i$. Assuming that $\phi_{xx}(x^i) > 0$, we get

$$\min_{\delta x} \left[\phi_x(x^i)\delta x + \frac{1}{2}\delta x^T \phi_{xx}(x^i)\delta x \right] \Rightarrow \delta x^i = -\phi_{xx}^{-1}(x^i)\phi_x^T(x^i), \tag{2.48}$$

giving

$$\Delta\phi(x^{i+1}) = -\frac{1}{2}\phi_x(x^i)\phi_{xx}^{-1}(x^i)\phi_x^T(x^i) + \mathcal{O}(||\delta x^i||^2). \tag{2.49}$$

Note that if ϕ is quadratic, the Newton–Raphson method converges to a minimum in one step.

Accelerated Gradient Methods

Since it is numerically inefficient to compute $\phi_{xx}(x^i)$, this second partial derivative can be estimated by constructing n independent directions from a sequence of gradients, $\phi_x^T(x^i)$, $i = 1, 2, \ldots, n$. For a quadratic function, this class of numerical optimization algorithms, called accelerated gradient methods, converges in n steps.

The most common of these methods are the *quasi-Newton methods*, so called because as the estimate of $\phi_{xx}(x^i)$, called the Hessian, approaches the actual value, the method approaches the Newton–Raphson method. The first and possibly most famous of these methods is still in popular use for solving unconstrained parameter optimization problems. It is known as the *Davidon–Fletcher–Powell* method [17] and dates from 1959. The method proceeds as a Newton–Raphson method where the inverse of $\phi_{xx}(x^i)$ is also estimated from the gradients and used as though it were the actual inverse of the Hessian. The most common implementation, described briefly here, uses a modified method of updating the estimate, known as the *Broyden–Fletcher–Goldfarb–Shanno*, or BFGS, update [9].

Let B_i be the estimate to $\phi_{xx}(x^i)$ at the ith iteration and g^i be the gradient $\phi_x(x^i)$. The method proceeds by computing the search direction s_i from

$$B_i s_i = -g^i \Rightarrow s_i = -B_i^{-1} g^i.$$

A one-dimensional search (using, possibly, the golden section search, Section 2.2.2) is performed along this direction, and the minimum found is taken as the next nominal set of parameters, x^{i+1}. The estimate of the inverse of the Hessian is then updated as $H_k = B_i^{-1}$,

$$\Delta g = g^{i+1} - g^i, \tag{2.50}$$

$$H_{i+1} = H_i - \frac{H_i \Delta g \Delta g^T H_i}{\Delta g^T H_i \Delta g} + \frac{s_i s_i^T}{\Delta g^T s_i}, \tag{2.51}$$

where $B_i > 0$. It can be shown that the method converges in n steps for a quadratic function and that for general functions, B_i converges to $\phi_{xx}(x^o)$ as $x^i \to x^o$ (assuming that $\phi_{xx}(x^o) > 0$).

For larger systems, a class of methods known as *conjugate gradient methods* requires less storage and also converges in n steps for quadratic functions. They converge less quickly for general functions, but since they do not require storing the Hessian estimate, they are preferred for very large systems. Many texts on these and other optimization methods (for example, [23] and [5]) give detailed discussions.

2.3 Minimization Subject to Constraints

The constrained parameter minimization problem is

$$\min_{x,u} \phi(x,u) \quad \text{subject to} \quad \psi(x,u) = 0, \tag{2.52}$$

where $x \in \mathbb{R}^n$, $u \in \mathbb{R}^m$, and $\psi(x,u)$ is a known n-dimensional vector of functions. Note that the cost criterion is minimized with respect to $n+m$ parameters. For ease of presentation the parameter vector is arbitrarily decomposed into two vectors (x,u). The point of this section is to convert a constrained problem to an unconstrained problem and then apply the results of necessity and sufficiency. We often choose

$\psi(x,u) = f(x,u) - c = 0$, where $f(x,u)$ are known n-dimensional functions and $c \in \mathbb{R}^n$ is a given vector, so that different levels of the constraint can be examined.

To illustrate the ideas and the methodology for obtaining necessary and sufficient conditions for optimality, we begin with a simple example, which is then extended to the general case. In this example the constrained optimization problem is transformed into an unconstrained problem, for which conditions for optimality were given in the previous sections. We will then relate this approach to the classical Lagrange multiplier method.

2.3.1 Simple Illustrative Example

Find the rectangle of maximum area inscribed in an ellipse defined by

$$f(x,u) = \frac{x^2}{a^2} + \frac{u^2}{b^2} = c, \tag{2.53}$$

where a, b, c are positive constants. The ellipse is shown in Figure 2.4 for $c = 1$. The area of a rectangle is the positive value of $(2x)(2u)$. The optimization problem is

$$\max_u (2x)(2u) = \min_u -4xu \stackrel{\triangle}{=} \min_u \phi(x,u) \tag{2.54}$$

subject to

$$(x,u) \in \Omega \stackrel{\triangle}{=} \{(x,u) | f(x,u) - c = 0\}, \tag{2.55}$$

where this becomes the area when $4xu$ is positive. It is assumed that $x \in \mathbb{R}$, $u \in \mathbb{R}$, $f : \mathbb{R}^2 \to \mathbb{R}$, and $\phi : \mathbb{R}^2 \to \mathbb{R}$.

The choice of u as the minimizing parameter where x satisfies the constraint is an arbitrary choice, and both x and u can be viewed as minimizing the cost criterion $\phi(x,u)$ and satisfying the constraint $\psi = f(x,u) - c = 0$. It is further assumed that f and ϕ are once continuously differentiable in each of their arguments. The main

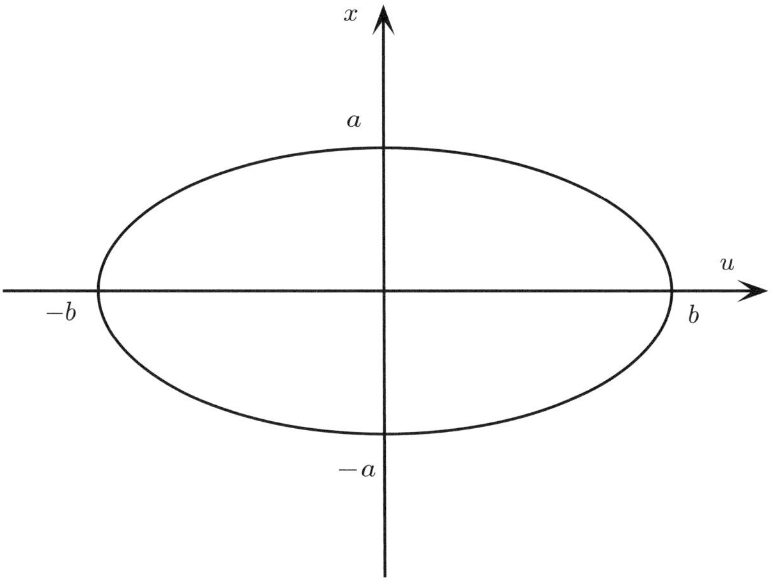

Figure 2.4: Ellipse definition.

difference between the constrained and unconstrained optimization problems is that in the constrained optimization problem, *the set* Ω *is not an open set.* Therefore, if (x^o, u^o) is an extremal, we cannot assert that $\phi(x^o, u^o) \leq \phi(x, u)$ for all (x, u) in an open set about (x^o, u^o) since any admissible variation must satisfy the constraint.

The procedure is to solve for x in terms of u to give an unconstrained problem for which u belongs to an open set. Some special considerations must be made on the function $\phi(x, u)$, which relates x and u. Note that for this problem, either $x^o \neq 0$ or $u^o \neq 0$, or both. Let $x^o \neq 0$ so that in a small region about $(x^o,\ u^o)$, i.e., for $x - x^o = \delta x$ and $u - u^o = \delta u$, $|\delta x| < \beta$, $|\delta u| < \epsilon$, the change in the constraint is

$$\begin{aligned} df &\stackrel{\triangle}{=} f(x^o + \delta x, u^o + \delta u) - f(x^o, u^o) \\ &= f_x(x^o, u^o)\delta x + f_u(x^o, u^o)\delta u + \mathcal{O}(d) = 0, \end{aligned} \tag{2.56}$$

where

$$d \stackrel{\triangle}{=} (\delta x^2 + \delta u^2)^{\frac{1}{2}} \quad \text{and} \quad \frac{\mathcal{O}(d)}{d} \to 0 \quad \text{as} \quad d \to 0. \tag{2.57}$$

Then, to first order,

$$\delta f = f_x(x^o, u^o)\delta x + f_u(x^o, u^o)\delta u = 0 \tag{2.58}$$

can be solved as

$$\delta x = -\left[f_x(x^o, u^o)\right]^{-1} f_u(x^o, u^o)\delta u \tag{2.59}$$

if $f_x(x^o, u^o) \neq 0$. This implies that $f(x, u) = c$ may be solved for x in terms of u. More precisely, if $f(x, u)$ is continuously differentiable and $f_x(x, u)$ is invertible, then the Implicit Function Theorem (see Appendix A.1.1) implies that there exists a rectangle $\bar{V}$ as $|x - x^o| < \beta$ and $|u - u^o| < \epsilon$, shown in Figure 2.5, such that

$$x = g(u), \tag{2.60}$$

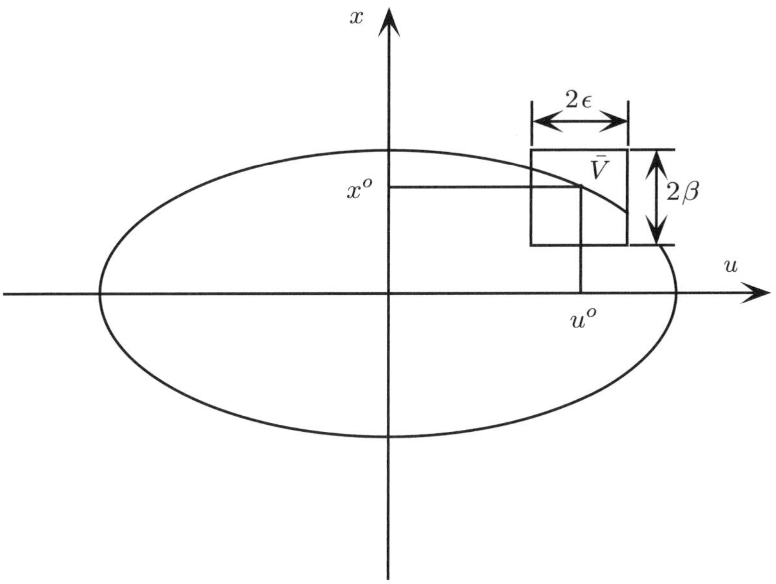

Figure 2.5: Definition of $\bar{V}$.

where $g(u)$ has continuous derivatives in $|u - u^o| < \epsilon$. Obtaining $g(u)$ *explicitly* may be quite difficult. However, this turns out *not* to be necessary since all that will be required is the *implicit* representation of $x = g(u)$, which is $f(x, u) = c$.

This implies that $x^o = g(u^o)$ and $f(g(u), u) = c$ whenever $|u - u^o| < \epsilon$. Since $(x^o, u^o) = (g(u^o), u^o)$ is an optimal point, it follows that u^o is the optimal solution for

$$\min_u \hat{\phi}(u) = \min_u \phi(g(u), u) \tag{2.61}$$

subject to $|u - u^o| < \epsilon$. Note that by explicitly eliminating the number of dependent variables x in terms of the independent variables u through the constraint, the objective function is now solved on an open set, where $\hat{\phi}(u)$ is continuously differentiable since ϕ and $g(u)$ are continuously differentiable. Therefore, our *unconstrained results* now apply. That is, using the chain rule,

$$\hat{\phi}_u(u^o) = \phi_x(x^o, u^o) g_u(u^o) + \phi_u(x^o, u^o) = 0. \tag{2.62}$$

We still need to determine $g_u(u)$. From $f(g(u), u) = c$ we obtain

$$f_x(x^o, u^o) g_u(u^o) + f_u(x^o, u^o) = 0, \tag{2.63}$$

$$\Rightarrow \quad g_u = -\frac{f_u}{f_x}. \tag{2.64}$$

The required first-order necessary condition is obtained by substituting (2.64) into (2.62) as

$$\phi_u - \phi_x \frac{f_u}{f_x} = 0 \quad \text{at} \quad x^o, u^o. \tag{2.65}$$

Note that g need not be determined. The optimal variables (x^o, u^o) are determined from two equations

$$\phi_u - \phi_x \frac{f_u}{f_x} = 0 \Rightarrow -4x + 4u\left(\frac{a^2}{2x}\frac{2u}{b^2}\right) = 0, \tag{2.66}$$

$$f(x, u) = c \Rightarrow \frac{x^2}{a^2} + \frac{u^2}{b^2} = c, \tag{2.67}$$

From (2.66) we obtain

$$x - \frac{u^2}{b^2}\frac{a^2}{x} = 0 \Rightarrow \frac{x^2}{a^2} - \frac{u^2}{b^2} = 0 \Rightarrow \frac{x^2}{a^2} = \frac{u^2}{b^2}. \tag{2.68}$$

Then, using (2.67) and (2.68), the extremal parameters

$$\frac{2u^2}{b^2} = c, \qquad \frac{2x^2}{a^2} = c, \tag{2.69}$$

$$\Rightarrow \quad u^o = \pm b\sqrt{\frac{c}{2}}, \qquad x^o = \pm a\sqrt{\frac{c}{2}}. \tag{2.70}$$

There are four extremal solutions, all representing the corners of the same rectangle. The minimum value is

$$\phi^o(c) = \hat{\phi}^o(c) = -2cab, \tag{2.71}$$

where the dependence of $\phi^o(c)$ on the constraint level c is explicit. The maximum value is $+2cab$.

First-Order Conditions for the Constrained Optimization Problem

We structure the necessary conditions given in Section 2.3.1 by defining a scalar λ as

$$\lambda \triangleq -\frac{\phi_x}{f_x}|_{(x^o,u^o)}. \tag{2.72}$$

Then (2.65) becomes

$$\phi_u = -\lambda f_u = -\lambda \psi_u, \tag{2.73}$$

and (2.72) becomes

$$\phi_x = -\lambda f_x = -\lambda \psi_x. \tag{2.74}$$

This means that at the optimal point, the gradient of ϕ is normal to the plane tangent to the constraint. This is depicted in Figure 2.6, where the tangent point is at the local minimum (u^o, x^o).

Finally, note that from (2.72)

$$\lambda = \frac{4u^o}{2x^o/a^2} = \frac{4b\sqrt{\frac{c}{2}}}{\frac{2}{a^2}(a\sqrt{\frac{c}{2}})} = 2ab, \tag{2.75}$$

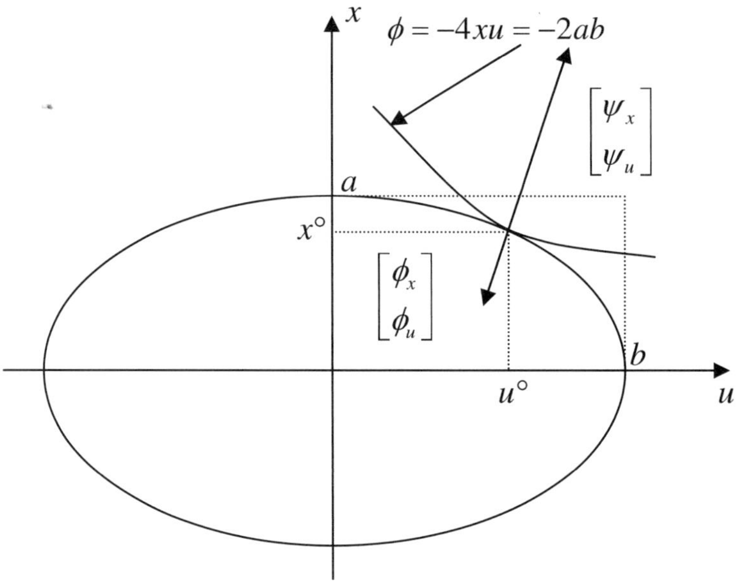

Figure 2.6: Definition of tangent plane.

which is related to ϕ of (2.71) by

$$\lambda = -\frac{\partial \phi^o(c)}{\partial c} = 2ab. \tag{2.76}$$

This shows that λ is an influence function relating a change in the optimal cost criterion to a change in the constraint level. We will later show that λ is related to the classical Lagrange multiplier.

Second-Order Necessary and Sufficient Conditions

In (2.65) the first-order condition for a scalar constraint is given. This along with the second variation give necessary and sufficient conditions for local optimality. Assuming that ϕ and f are twice differentiable, then so are g and $\hat{\phi}(u)$. Since u^o lies in an open set, the second-order necessary condition obtained for the unconstrained optimization problem applies here as well. Therefore,

$$\hat{\phi}_{uu}(u^o) = \phi_{uu} + 2\phi_{xu}g_u + \phi_x g_{uu} + \phi_{xx}g_u^2 \geq 0. \tag{2.77}$$

To determine g_u and g_{uu} we expand $f(g(u),u) \triangleq \hat{f}(u) = c$ about u^o and note that the coefficients of δu and δu^2 must be zero,

$$\hat{f}(u^o + \delta u) = \hat{f}(u^o) + \hat{f}_u(u^o)\delta u + \frac{1}{2}\hat{f}_{uu}(u^o)\delta u^2 + \cdots = c, \tag{2.78}$$

where

$$\hat{f}_u = f_x g_u + f_u = 0 \Rightarrow g_u = -\frac{f_u}{f_x}, \tag{2.79}$$

$$\hat{f}_{uu} = f_{xx}g_u^2 + f_x g_{uu} + 2f_{xu}g_u + f_{uu} = 0, \tag{2.80}$$

$$\Rightarrow g_{uu} = -\frac{1}{f_x}\left[f_{uu} - 2f_{xu}\frac{f_u}{f_x} + f_{xx}\left(\frac{f_u}{f_x}\right)^2\right]. \tag{2.81}$$

Substitution into $\hat{\phi}_{uu}$ given in (2.77) produces the desired condition

$$\left[\phi_{uu} - \frac{\phi_x}{f_x}f_{uu}\right] - 2\left[\phi_{xu} - \frac{\phi_x}{f_x}f_{xu}\right]\frac{f_u}{f_x} + \left[\phi_{xx} - \frac{\phi_x}{f_x}f_{xx}\right]\left(\frac{f_u}{f_x}\right)^2 \geq 0. \tag{2.82}$$

By identifying $\lambda = -\frac{\phi_x}{f_x}$, then

$$\hat{\phi}_{uu}(u^o) = [\phi_{uu} + \lambda f_{uu}] - 2[\phi_{xu} + \lambda f_{xu}]\frac{f_u}{f_x} + [\phi_{xx} + \lambda f_{xx}]\left(\frac{f_u}{f_x}\right)^2 \geq 0. \tag{2.83}$$

A necessary condition for a local minimum is that $\hat{\phi}_{uu}(u^o) \geq 0$. The sufficient conditions for a local minimum are $\hat{\phi}_u(u^o) = 0$ and $\hat{\phi}_{uu}(u^o) > 0$.

We derive the first- and second-order necessary and sufficient conditions by an alternate method, which has sufficient generality that it is used to generate these necessary and sufficient conditions in the most general case. In particular, for this two-parameter problem we introduce the Lagrange multiplier method. The cost criterion is augmented by the constraint by the Lagrange multiplier as $H \triangleq \phi(x,u) + \lambda(f(x,u) - c)$. Expanding the augmented cost H in a Taylor series to second order,

$$H(x^o + \delta x, u^o + \delta u, \lambda + \delta\lambda) - H(x^o, u^o, \lambda) \cong H_x\delta x + H_u\delta u + H_\lambda\delta\lambda$$
$$+\frac{1}{2}[\delta x \quad \delta u \quad \delta\lambda]\begin{bmatrix} H_{xx} & H_{xu} & H_{x\lambda} \\ H_{ux} & H_{uu} & H_{u\lambda} \\ H_{\lambda x} & H_{\lambda u} & H_{\lambda\lambda} \end{bmatrix}\begin{bmatrix} \delta x \\ \delta u \\ \delta\lambda \end{bmatrix} \triangleq \frac{1}{2}\delta^2 H, \tag{2.84}$$

where from (2.73) $H_u = 0$, from (2.74) $H_x = 0$ and $H_\lambda = f(x,u) - c = 0$. There is *no* requirement that the second-order term be positive semidefinite for arbitrary variations in (x, u, λ). Intuitively, the requirement is that the function ϕ takes on a minimum value on the tangent plane of the constraint. This is done by using the relation between δx and δu of

$$\delta x = -\frac{f_u(x^o, u^o)}{f_x(x^o, u^o)} \delta u. \tag{2.85}$$

If this is substituted into the quadratic form in (2.84), the quadratic form reduces to (note $H_{\lambda\lambda} \equiv 0$)

$$\delta^2 H = \delta u \left[H_{xx} \left(\frac{f_u}{f_x} \right)^2 - 2H_{xu} \frac{f_u}{f_x} + H_{uu} \right] \delta u \geq 0, \tag{2.86}$$

where *the coefficient of* $\delta\lambda$ *becomes identically zero*, i.e., $f_x \delta x + f_u \delta u = 0$. The coefficient of the quadratic in (2.86) is identical to (2.83).

For the particular example of finding the largest rectangle in an ellipse, the second variation of (2.83) is verified as $\hat{\phi}_{uu}(u^o) = 16a/b > 0$, ensuring that ϕ is a locally constrained maximum at u^o.

2.3.2 General Case: Functions of n-Variables

Theorem 2.3.1 *Let* $f_i : \mathbb{R}^{n+m} \to \mathbb{R}$, $i = 1, \ldots, n$, *be* n *continuously differentiable constraints and* $\phi : \mathbb{R}^{n+m} \to \mathbb{R}$ *be the continuously differentiable performance index. Let* $x^o \in \mathbb{R}^n$ *and* $u^o \in \mathbb{R}^m$ *be the optimal variables of the problem*

$$\phi^o = \min_{x,u} \phi(x, u) \tag{2.87}$$

subject to $f_i(x, u) = c_i$, $i = 1, \ldots, n$, *or* $f(x, u) = c$. *Suppose that at* (x^o, u^o) *the* $n \times n$ *matrix* $f_x(x^o, u^o)$ *is nonsingular, then there exists a vector* $\lambda \in \mathbb{R}^n$ *such that*

$$\phi_x(x^o, u^o) = -\lambda^T f_x(x^o, u^o), \tag{2.88}$$

$$\phi_u(x^o, u^o) = -\lambda^T f_u(x^o, u^o). \tag{2.89}$$

Furthermore, if $(x^o(c), u^o(c))$ *are once continuously differentiable functions of* $c = [c_1, \dots, c_n]^T$, *then* $\phi^o(c)$ *is a differentiable function of* c *and*

$$\lambda^T = -\frac{\partial \phi^o(c)}{\partial c}. \tag{2.90}$$

Remark 2.3.1 *We choose* $\psi(x, u) = f(x, u) - c = 0$ *without loss of generality so that different levels of the constraint* c *can be examined and related to* ϕ^o *as given in* (2.90).

Proof: Since $f_x(x^o, u^o)$ is nonsingular, by the Implicit Function Theorem (see section A.1.1) there exists an $\epsilon > 0$, an open set $V \in \mathbb{R}^{n+m}$ containing (x^o, u^o), and a differentiable function $g : \mathcal{U} \to \mathbb{R}^n$, where $\mathcal{U} = [u : \|u - u^o\| < \epsilon]$.[3] This means that

$$f(x, u) = c, \quad [x^T, u^T] \in V, \tag{2.91}$$

implies that

$$x = g(u) \quad \text{for} \quad u \in \mathcal{U}. \tag{2.92}$$

Furthermore, $g(u)$ has a continuous derivative for $u \in \mathcal{U}$.

Since $(x^o, u^o) = (g(u^o), u^o)$ is optimal, it follows that u^o is an optimal variable for a new optimization problem defined by

$$\min_u \phi(g(u), u) = \min_u \hat{\phi}(u) \quad \text{subject to} \quad u \in \mathcal{U}. \tag{2.93}$$

$\mathcal{U}$ is an open subset of $\mathbb{R}^m$ and $\hat{\phi}$ is a differentiable function on $\mathcal{U}$, since ϕ and g are differentiable. Therefore, Theorem 2.2.2 is applicable, and by the chain rule

$$\hat{\phi}_u(u^o) = \phi_x g_u + \phi_u|_{u=u^o, x=g(u^o)} = 0. \tag{2.94}$$

[3] $\mathcal{U}$ is the set of points u that lie in the ball defined by $\|u - u^o\| < \epsilon$.

Furthermore, $f(g(u), u) = c$ for all $u \in \mathcal{U}$. This means that all derivatives of $f(g(u), u)$ are zero, in particular the first derivative evaluated at (x^o, u^o):

$$f_x g_u + f_u = 0. \tag{2.95}$$

Again, since the matrix function $f_x(x^o, u^o)$ is nonsingular, we can evaluate g_u as

$$g_u = -f_x^{-1} f_u. \tag{2.96}$$

Substitution of g_u into (2.94) gives

$$\left[\phi_u - \phi_x f_x^{-1} f_u\right]_{(x^o,u^o)} = 0. \tag{2.97}$$

Let us now define the n-vector λ as

$$\lambda^T = -\phi_x f_x^{-1}|_{(x^o,u^o)}. \tag{2.98}$$

Then (2.97) and (2.98) can be written as

$$[\phi_x, \phi_u] = -\lambda^T \left[f_x, f_u\right]. \tag{2.99}$$

Now we will show that $\lambda^T = -\phi_c^o(c)$. Since by assumption $f(x, u)$ and $(x^o(c), u^o(c))$ are continuously differentiable, it follows that in a neighborhood of c, f_x is nonsingular. Then

$$f(x^o(c), u^o(c)) \quad = \quad c, \tag{2.100}$$

$$\phi_u - \phi_x f_x^{-1} f_u \quad = \quad 0 \tag{2.101}$$

using the first-order condition. By differentiating $\phi^o(c) = \phi(u^o, x^o)$,

$$\phi_c^o = \phi_x x_c^o + \phi_u u_c^o. \tag{2.102}$$

Differentiating $f(x^o(c), u^o(c)) = c$ gives

$$f_x x_c^o + f_u u_c^o = I \Rightarrow x_c^o + f_x^{-1} f_u u_c^o = f_x^{-1}. \tag{2.103}$$

Multiplying by ϕ_x gives

$$\phi_x x_c^o + \phi_x f_x^{-1} f_u u_c^o = \phi_x f_x^{-1} = -\lambda^T. \tag{2.104}$$

Using the first-order condition of (2.89) gives the desired result

$$\phi_c^o = -\lambda^T. \tag{2.105}$$

∎

Remark 2.3.2 *Equation* (2.99) *shows that the gradient of the cost function* $[\phi_x, \phi_u]$ *is orthogonal to the tangent plane of the constraint at* (x^o, u^o). *Since* $[f_x, f_u]$ *form* n *independent vectors (because* f_x *is nonsingular), then the tangent plane is described by the set of vectors* h *such that* $[f_x, f_u]h = 0$.

The set of vectors which are orthogonal to this tangent surface is any linear combination of the gradient $[f_x, f_u]$. *In particular, if* $\lambda^T[f_x, f_u]$ *is such that*

$$[\phi_x, \phi_u] = -\lambda^T[f_x, f_u], \tag{2.106}$$

then $[\phi_x, \phi_u]$ *is orthogonal to the tangent surface.*

Lagrange Multiplier Approach

Identical necessary conditions to those obtained above are derived formally by the Lagrange multiplier approach. By adjoining the constraint $f = c$ to the cost function ϕ with an undetermined n-vector Lagrange multiplier λ, a function $H(x, u, \lambda)$ is defined as[4]

$$H(x, u, \lambda) \triangleq \phi(x, u) + \lambda^T(f(x, u) - c), \tag{2.107}$$

and we construct an unconstrained optimization problem in the $2n + m$ variables x, u, λ. Therefore, we look for the *extremal point of* H with respect to x, u, and λ,

[4]Note that $H(x, u, \lambda) = \phi(x, u)$ when the constraint is satisfied.

where x, u, and λ are considered free and can be arbitrarily varied within some small open set containing x^o, u^o, and λ. From our unconstrained optimization results we have

$$H_x = \phi_x + \lambda^T f_x = 0, \tag{2.108}$$

$$H_u = \phi_u + \lambda^T f_u = 0, \tag{2.109}$$

$$H_\lambda = f(x^o, u^o) - c = 0. \tag{2.110}$$

This gives us $2n+m$ equations in $2n+m$ unknowns x, u, λ. Note that satisfaction of the constraint is now satisfaction of the necessary condition (2.110).

2.3.3 Constrained Parameter Optimization: An Algorithmic Approach

In this section we extend the steepest descent method of Section 2.2.5 to include equality constraints. The procedure suggested is to first satisfy the constraint, i.e., constraint restoration. Then, a gradient associated with changes in the cost criterion along the tangent plane to the constraint manifold is constructed. This is done by forming a projector that annihilates any component of the gradient of the cost criterion in the direction of the gradient of the constraint function. Since these gradients are determined from the first-order term in a Taylor series of the cost criterion and the constraint functions, the steps used in the iteration process must be sufficiently small to preserve the validity of this assumed linearity.

Suppose $y \triangleq [x^T, u^T]^T$. The parameter optimization problem is

$$\min_y \phi(y) \quad \text{subject to} \quad f(y) = c, \tag{2.111}$$

where ϕ and f are assumed to be sufficiently smooth so that for small changes in y away from some nominal value y^i, ϕ and f can be approximated by the first term of a Taylor series about y^i as $(\delta y = y - y^i)$,

$$\delta\phi \cong \phi_y \delta y, \tag{2.112}$$

$$\delta f \cong f_y \delta y. \tag{2.113}$$

In the following we describe a numerical optimization algorithm composed of a constraint restoration step followed in turn by a minimization step. Although these steps can be combined, they are separated here for pedagogical reasons.

Constraint Restoration

Since $f = c$ describes a manifold in y space and assuming y^i is a point *not* on $f = c$, from (2.113) a change in the constraint level is related to a change in y. To move in the direction of constraint satisfaction choose δy as

$$\delta y = f_y^T (f_y f_y^T)^{-1} \delta f, \tag{2.114}$$

where the choice of $\delta f = -\epsilon\psi = \epsilon(c - f(y^i))$ for small $\epsilon > 0$ forms an iterative step of driving f to c. Note that δy in (2.114) is a least-squares solution to (2.113) where f_y is full rank. At the end of each iteration to satisfy the constraint, set $y^i = y^i + \delta y$. The iteration sequence stops when for $\epsilon_1 > 0$, $|c - f| < \epsilon_1 \ll 1$.

Constrained Minimization

Since $f = c$ describes a manifold in y space and assuming y^i is a point on $f = c$, then f_y is perpendicular to the tangent plane of $f = c$ at y^i. To ensure that changes in the cost $\phi(y)$ are made only in the tangent plane, so that the constraint will not be violated (to first order), define the projection operator as

$$P = I - f_y^T (f_y f_y^T)^{-1} f_y, \tag{2.115}$$

which has the properties that

$$PP = P, \quad Pf_y^T = 0, \quad P = P^T. \tag{2.116}$$

Therefore, the projection operator will annihilate components of a vector along f_y^T. The object is to use this projector to ensure that if changes are made in improving the cost, they are made in only the tangent line to the constraint. See Figure 2.7 for a geometrical description where $y \triangleq [x^T, u^T]^T$ and f, x, u are scalars.

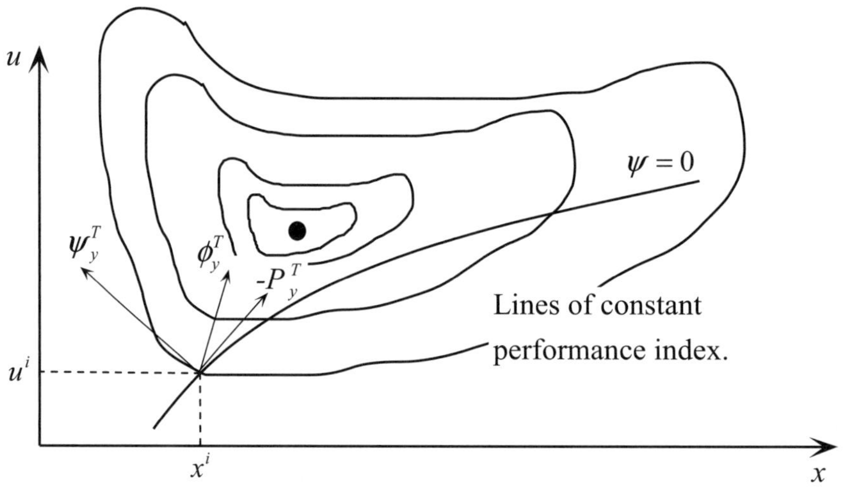

Figure 2.7: Geometrical description of parameter optimization problem.

This projected gradient is constructed by choosing the control changes in the steepest descent direction, while tangential to the constraint $\psi(x, u) = f(x, u) - c = 0$, i.e.,

$$\delta y = -\epsilon P \phi_y^T, \tag{2.117}$$

where again ϵ is a positive number chosen small so as not to violate the assumed linearity. With this choice of δy, the cost criterion change to first order is

$$\delta\phi = -\epsilon \phi_y P \phi_y^T = -\epsilon \phi_y P P^T \phi_y^T = -\epsilon \|\phi_y P\|^2 \tag{2.118}$$

(since $PP^T = P$). Note that the constraint is satisfied to first order ($\|\delta y\| < \epsilon \ll 1$)

$$\Delta f = \delta f + \mathcal{O}(\|\delta y\|), \tag{2.119}$$

where $\delta f = f_y \delta y = -\epsilon f_y P \phi_y^T = 0$. The second-order constraint violation is then restored by going back to the constraint restoration step. This iterative process between constraint restoration and constrained minimization is continued until the stationary necessary conditions

$$P\phi_y^T = 0, \quad f = c \tag{2.120}$$

are met. Note that the constraint restoration and optimization steps can be combined given the assumed linearity. This optimization algorithm is called steepest descent optimization with constraints.

The Lagrange multiplier technique is consistent with the results of (2.120). The n-vector Lagrange multiplier λ is now shown to contribute to the structure of the projector. If the constraint variation δf is adjoined to $\delta\phi$ by the Lagrange multiplier λ in (2.113), the augmented cost variation $\delta\bar{\phi}$ is

$$\delta\bar{\phi} = (\phi_y + \lambda^T f_y)\delta y. \tag{2.121}$$

If δy is chosen as

$$\delta y = -\epsilon(\phi_y + \lambda^T f_y)^T, \tag{2.122}$$

then by the usual arguments, a minimum occurs for the augmented cost when

$$\phi_y + \lambda^T f_y = 0, \tag{2.123}$$

where $f(y^i) - c = 0$ and $f_y \delta y = 0$. Postmultiplying (2.123) by f_y^T and solving for λ at (x^o, u^o) results in

$$\lambda^T = -\phi_y f_y^T (f_y f_y^T)^{-1}. \tag{2.124}$$

Substituting (2.124) back into (2.123) results in

$$\phi_y - \phi_y f_y^T (f_y f_y^T)^{-1} f_y = \phi_y \left[I - f_y^T (f_y f_y^T)^{-1} f_y\right] = \phi_y P = 0, \tag{2.125}$$

which is just the first condition of (2.120).

The constraint projection and restoration method described here is an effective method for numerical solution of minimization problems subject to equality constraints. Several other methods are also in common use, with many sharing significant ideas.

All such methods are subject to a number of difficulties in actual implementation. These are beyond the scope of this text, but the interested reader may see [23], [5], and [36] for more information.

2.3.4 General Form of the Second Variation

Assume that ϕ and f are twice differentiable. Since f is twice differentiable, so is g and, therefore, $\hat{\phi}(u)$. Whereas u^o lies in an open set, the general second-order necessary condition for an unconstrained optimization problem applies. Producing the inequality by the procedure of the previous section is laborious. Rather, we use the equivalent Lagrange multiplier approach, where

$$H = \phi + \lambda^T \psi. \tag{2.126}$$

(If $\psi = f(x, u) - c = 0$, then $H = \phi + \lambda^T (f - c)$.) Then, expanding the augmented cost to second order, assuming first-order necessary conditions hold,

$$\begin{aligned} H(x^o + \delta x, u^o + \delta u, \lambda + \delta\lambda) - H(x^o, u^o, \lambda) &= \frac{1}{2}\delta^2 H + \mathcal{O}(d^2) \\ = \frac{1}{2} \left[\, \delta x^T \quad \delta u^T \quad \delta\lambda^T \,\right] \begin{bmatrix} H_{xx} & H_{xu} & H_{x\lambda} \\ H_{ux} & H_{uu} & H_{u\lambda} \\ H_{\lambda x} & H_{\lambda u} & H_{\lambda\lambda} \end{bmatrix} &\begin{bmatrix} \delta x \\ \delta u \\ \delta\lambda \end{bmatrix} + \mathcal{O}(d^2), \end{aligned} \tag{2.127}$$

where the first-order condition $\delta H = 0$ ($H_x = 0$, $H_u = 0$ and $H_\lambda = f - c = 0$) is used and $d = \|\delta x^T, \delta u^T, \delta\lambda^T\|$. From $x^o = g(u^o)$ and its properties in $(x, u) \in V$ (see the Implicit Function Theorem, Section A.1.1),

$$\delta x = g_u \delta u + \mathcal{O}(\|\delta u\|), \tag{2.128}$$

where $f(g(u), u) \triangleq \hat{f}(u) = c$ requires that all its derivatives be zero. In particular,

$$\hat{f}_u = H_{\lambda x} g_u + H_{\lambda u} = f_x g_u + f_u = 0 \Rightarrow g_u = -f_x^{-1} f_u. \tag{2.129}$$

Using (2.128) and (2.129) in (2.127), the second variation reduces to

$$\delta^2 H = \delta u^T \left[g_u^T H_{xx} g_u + H_{ux} g_u + g_u^T H_{xu} + H_{uu} \right] \delta u \geq 0. \tag{2.130}$$

Therefore, the necessary condition for local optimality is

$$\begin{bmatrix} g_u^T & I \end{bmatrix} \begin{bmatrix} H_{xx} & H_{xu} \\ H_{ux} & H_{uu} \end{bmatrix} \begin{bmatrix} g_u \\ I \end{bmatrix} \geq 0, \tag{2.131}$$

and the sufficiency condition is

$$\begin{bmatrix} g_u^T & I \end{bmatrix} \begin{bmatrix} H_{xx} & H_{xu} \\ H_{ux} & H_{uu} \end{bmatrix} \begin{bmatrix} g_u \\ I \end{bmatrix} > 0, \tag{2.132}$$

along with the first-order conditions.

Note 2.3.1 *Again, the coefficients associated with $\delta\lambda$ are zero.*

Note 2.3.2 *The definiteness conditions are only in an m-dimensional subspace associated with the tangent plane of the constraints evaluated at (x^o, u^o).*

2.3.5 Inequality Constraints: Functions of 2-Variables

An approach for handling optimization problems with inequality constraints is to convert the inequality constraint to an equality constraint by using a device called slack variables. Once the problem is in this form, the previous necessary conditions are applicable. We present the 2-variable optimization problem for simplicity, but the extension to n dimensions is straightforward.

Theorem 2.3.2 *Let $\phi : \mathbb{R}^2 \to \mathbb{R}$ and $\theta : \mathbb{R}^2 \to \mathbb{R}$ be twice continuously differentiable. Let (x^o, u^o) be the optimal variables of the problem*

$$\phi^o = \min_{x,u} \phi(x,u) \quad \textit{subject to} \quad \theta(x,u) \leq 0, \tag{2.133}$$

where if $\theta(x^o, u^o) = 0$, then $\theta_x \neq 0$. Then there exists a scalar $\nu \geq 0$ such that

$$(\phi_x, \phi_u) = -\nu(\theta_x, \theta_u). \tag{2.134}$$

Remark 2.3.3 *If x^o, u^o lies in the interior of the constraint, then $\nu = 0$ and the necessary conditions become those of an unconstrained minimization problem, i.e., $(\phi_x, \phi_u) = 0$. If x^o, u^o lies on the boundary of the constraint, $\nu > 0$ and the necessary conditions of* (2.134) *hold.*

Proof: We first convert the inequality constraint into an equality constraint by introducing a slack variable α such that

$$\theta(x,u) = -\alpha^2. \tag{2.135}$$

For any real value of $\alpha \in \mathbb{R}$ the inequality constraint is satisfied. An equivalent problem is

$$\phi^o = \min_{x,u} \phi(x,u) \quad \text{subject to} \quad \theta(x,u) + \alpha^2 = 0, \tag{2.136}$$

which was considered in the last section.

For simplicity, we use the Lagrange multiplier approach here. Adjoin the constraint $\theta + \alpha^2 = 0$ to the cost function ϕ by an undetermined multiplier ν as

$$H = \phi(x, u) + \nu(\theta(x, u) + \alpha^2). \tag{2.137}$$

We look for the extremal point for H with respect to (x, u, α, ν). From our unconstrained optimization results we have

$$H_x = \phi_x + \nu^o \theta_x = 0 \quad \text{at } (x^o, u^o), \tag{2.138}$$

$$H_u = \phi_u + \nu^o \theta_u = 0 \quad \text{at } (x^o, u^o), \tag{2.139}$$

$$H_\nu = \theta(x^o, u^o) + \alpha^{o2} = 0, \tag{2.140}$$

$$H_\alpha = 2\nu^o \alpha^o = 0. \tag{2.141}$$

This gives four equations in four unknowns x, u, ν, α. From (2.138) and (2.139) we obtain the condition shown in (2.134).

The objective now is to show that $\nu \geq 0$. If $\alpha^o \neq 0$, then $\nu^o = 0$ off the boundary of the constraint (in the admissible interior). If $\alpha^o = 0$, then the optimal solution is on the constraint boundary where $\nu^o \neq 0$. To determine if $\nu^o \geq 0$, the second variation is used as

$$\delta^2 H = [\,\delta x \quad \delta u \quad \delta\alpha \quad \delta\nu\,] \begin{bmatrix} H_{xx} & H_{xu} & H_{x\alpha} & H_{x\nu} \\ H_{ux} & H_{uu} & H_{u\alpha} & H_{u\nu} \\ H_{\alpha x} & H_{\alpha u} & H_{\alpha\alpha} & H_{\alpha\nu} \\ H_{\nu x} & H_{\nu u} & H_{\nu\alpha} & H_{\nu\nu} \end{bmatrix} \begin{bmatrix} \delta x \\ \delta u \\ \delta\alpha \\ \delta\nu \end{bmatrix}, \tag{2.142}$$

where the variation of the constraint is used to determine δx in terms of δu and $\delta\alpha$ as

$$\theta_x \delta x + \theta_u \delta_u + 2\alpha\delta\alpha = 0. \tag{2.143}$$

However, on the constraint, $\alpha^o = 0$. Since by assumption $\theta_x \neq 0$,

$$\delta x = -\frac{\theta_u}{\theta_x}\delta u. \tag{2.144}$$

In addition,

$$\begin{aligned} H_{x\alpha} &= 0, \\ H_{u\alpha} &= 0, \\ H_{\nu\nu} &= 0, \\ H_{\alpha\alpha} &= 2\nu^o, \\ H_{\nu\alpha} &= 2\alpha^o = 0. \end{aligned} \tag{2.145}$$

Using (2.144) and (2.145) in the second variation, (2.142) reduces to

$$\delta^2 H = [\,\delta u \quad \delta\alpha\,] \begin{bmatrix} \left[H_{uu} - 2H_{ux}\frac{\theta_u}{\theta_x} + H_{xx}\left(\frac{\theta_u}{\theta_x}\right)^2 \right] & 0 \\ 0 & 2\nu \end{bmatrix} \begin{bmatrix} \delta u \\ \delta\alpha \end{bmatrix} \geq 0, \tag{2.146}$$

and then $\delta^2 H$ is positive semidefinite if

$$H_{uu} - 2H_{ux}\frac{\theta_u}{\theta_x} + H_{xx}\left(\frac{\theta_u}{\theta_x}\right)^2 \geq 0, \tag{2.147}$$

$$\nu \geq 0. \tag{2.148}$$

∎

Note: The second variation given above is only for the case in which the optimal variables lie on the boundary. If the optimal point lies in the interior of the constraint, then the unconstrained results apply.

This simple example can be generalized to the case with n inequality constraints. There are many fine points in the extension of this theory. For example, if all the inequality constraints are feasible at or below zero, then under certain conditions the gradient of the cost criterion is contained at the minimum to be in a cone constructed from the gradients of the active constraint functions (i.e., $\alpha = 0$ in the above two-variable case). This notion is implied by the Kuhn–Tucker theorem [33]. In this chapter, we have attempted only to give an introduction that illuminates the principles of optimization theory and the concepts that will be used in following chapters.

Problems

1. A tin can manufacturer wants to find the dimensions of a cylindrical can (closed top and bottom) such that, for a given amount of tin, the volume of the can is a maximum. If the thickness of the tin stock is constant, a given amount of tin implies a given surface area of the can. Use height and radius as variables and use a Lagrange multiplier.

2. Determine the point x_1, x_2 at which the function

$$\phi = x_1 + x_2$$

is a minimum, subject to the constraint

$$x_1^2 + x_1 x_2 + x_2^2 = 1.$$

3. Minimize the performance index

$$\phi = \frac{1}{2}(x^2 + y^2 + z^2)$$

subject to the constraints

$$x + 2y + 3z = 10,$$
$$x - y + 2z = 1.$$

Show that

$$x = \frac{19}{59}, \quad y = \frac{146}{59}, \quad z = \frac{93}{59}, \quad \lambda_1 = \frac{-55}{59}, \quad \lambda_2 = \frac{36}{59}.$$

4. Minimize the performance index

$$\phi = \frac{1}{2}(x^2 + y^2 + z^2)$$

subject to the constraint

$$x + 2y - 3z - 7 = 0.$$

5. Minimize the performance index

$$\phi = x - y + 2z$$

subject to the constraint

$$x^2 + y^2 + z^2 = 2.$$

6. Maximize the performance index

$$\phi = x_1 x_2$$

subject to the constraint

$$x_1 + x_2 - 1 = 0.$$

7. Minimize the performance index

$$\phi = -x_1 x_2 + x_2 x_3 + x_3 x_1$$

subject to the constraint

$$x_1 + x_2 - x_3 + 1 = 0$$

8. Minimize the performance index

$$\phi = \sqrt{4 - 3x^2}$$

subject to the constraint

$$-1 \leq x \leq 1.$$

9. Maximize the performance index

$$\phi = xu$$

subject to the inequality constraint

$$x + u \leq 1.$$

10. (a) State the necessary and sufficient conditions and underlying assumptions for x^*, u^* to be locally minimizing for the problem of minimizing

$$\phi = \phi(u, x) \qquad x \in \mathbb{R}^n, \quad u \in \mathbb{R}^m,$$

and subject to

$$Ax + Bu = C.$$

(b) Find the extremals of

$$\phi = e^{x_1^2 + x_2}$$

subject to

$$x_1^2 + x_2^2 = \frac{1}{2}.$$

11. (a) In the two-dimensional xt-plane, determine the extremal curve of stationary length which starts on the circle $x^2 + t^2 - 1 = 0$ and terminates on the line $t = T = 2$.

(b) Solve problem (a) but consider that the termination is on the line $-x + t = 2\sqrt{2}$.

Note: Parts (a) and (b) are *not* to be solved by inspection.

CHAPTER 3

Optimization of Dynamic Systems with General Performance Criteria

3.1 Introduction

In accordance with the theme of this book outlined in Chapter 1, we use linear algebra, elementary differential equation theory, and the definition of the derivative to derive conditions that are satisfied by a control function which optimizes the behavior of a dynamic system relative to a specified performance criterion. In other words, we derive necessary conditions and also a sufficient condition for the optimality of a given control function.

In Section 3.2 we begin with the control of a linear dynamic system relative to a general performance criterion. Restricting attention to a linear system and introducing the notion of weak control perturbations allows an easy derivation of a weak form of the first-order necessary conditions. Then we extend these necessary conditions to nonlinear systems with the aid of a theorem by Bliss [6] on the differentiability of the solution of an ordinary differential equation with respect to a parameter. Next, we comment upon the two-point boundary-value problem based on these necessary conditions. We then introduce the notion of strong control perturbations, which allows the derivation of a stronger form of the first-order necessary conditions, which are referred to as Pontryagin's Principle [38]. This result is further strengthened upon the introduction of control variable constraints. After having observed that Pontryagin's Principle is only a necessary condition for optimality, we introduce the Hamilton–Jacobi–Bellman (H-J-B) equation and provide a general sufficient condition for optimality. The dependent variable of the H-J-B partial differential equation is the optimal value function, which is the value of the cost criterion using the optimal control. Using the H-J-B equation, we relate the derivative of the optimal value function to Pontryagin's Lagrange multipliers. Then we derive the H-J-B equation on the assumption that the optimal value function exists and is once continuously differentiable. Finally, we treat the case of unspecified final time and derive an additional necessary condition, called the transversality condition. We illustrate, where necessary, the conditions that we develop in this chapter by working out several examples.

Remark 3.1.1 *Throughout this book, time (or the variable t) is considered the independent variable. This need not always be the case. In fact, the choice of what constitutes a state, a control, and a "running variable" can drastically alter the ease with which a problem may be solved. For example, in a rocket launch, the energy of the vehicle (kinetic plus potential) can be considered as a state, a control, or the independent variable. The choice of which depends on the specifics of the problem at hand. Since once those items are chosen the notation becomes a matter of choice, we will stick to calling the states x, the controls u, and the independent variable t, and the problems in this book are laid out in that notation.*

3.2 Linear Dynamic Systems with General Performance Criterion

The linear dynamic system to be controlled is described by the vector linear differential equation

$$\begin{aligned} \dot{x}(t) &= A(t)x(t) + B(t)u(t), \\ x(t_0) &= x_0, \end{aligned} \tag{3.1}$$

where $x(\cdot)$ and $u(\cdot)$ are, respectively, n and m vector functions of time t, and where $A(\cdot)$ and $B(\cdot)$ are $n \times n$ and $n \times m$ matrix functions of time t. The initial condition at time $t = t_0$ for (3.1) is x_0. We make the following assumptions.

Assumption 3.2.1 *The elements $a_{ij}(\cdot)$ and $b_{kl}(\cdot)$ of $A(\cdot)$ and $B(\cdot)$ are continuous functions of t on the interval $[t_0, t_f]$, $t_f > t_0$.*

Assumption 3.2.2 *The control function $u(\cdot)$ is drawn from the set $\mathcal{U}$ of piecewise continuous m-vector functions of t on the interval $[t_0, t_f]$.*

The optimal control problem is to find a control function $u^o(\cdot)$ which minimizes the performance criterion

$$J\left(u(\cdot); x_0\right) = \phi\left(x(t_f)\right) + \int_{t_0}^{t_f} L\left(x(t), u(t), t\right) dt. \tag{3.2}$$

Here L, the Lagrangian, and ϕ are scalar functions of their arguments, and we make the following assumptions concerning these functions.

Note 3.2.1 *The notation $(\cdot)$, as used in $L(\cdot)$, is used to denote the functional form of L.*

Assumption 3.2.3 *The scalar function $L(\cdot,\cdot,\cdot)$ is once continuously differentiable in x and u and is continuous in t on $[t_0, t_f]$.*

Assumption 3.2.4 *The scalar function $\phi(\cdot)$ is once continuously differentiable in x.*

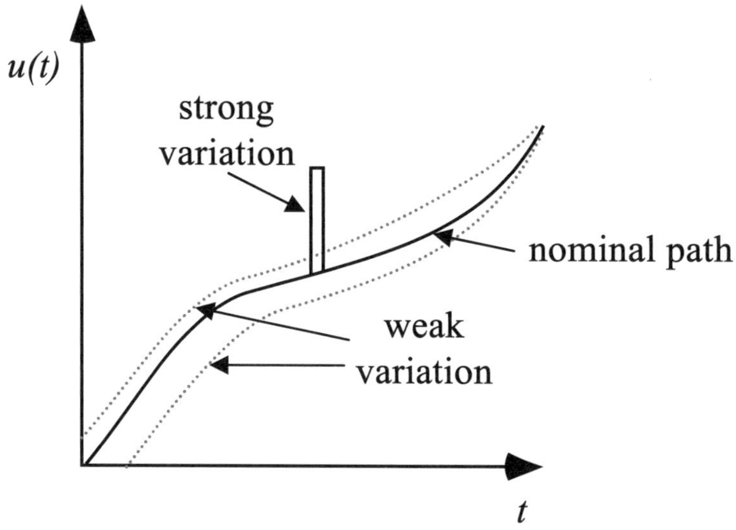

Figure 3.1: Depiction of weak and strong variations.

We now suppose that there is a piecewise continuous control function $u^o(\cdot)$ that minimizes (3.2), and we derive first-order conditions which this control will satisfy. In the derivation to come we make use of the properties of (3.1) developed in the next section.

Remark 3.2.1 *In this chapter, two types of variations (or perturbations) are considered: strong and weak variations. These are depicted graphically in Figure* 3.1. *Note that the strong variation is characterized by large variations over a very short interval. Contrast this with the weak variations, which are small perturbations over a large time interval. Also note that under the correct conditions (to be introduced* *in this chapter), both weak and strong variations in the control can produce "small" state variations.*

3.2.1 Linear Ordinary Differential Equation

Under the assumptions made in Section 3.2 it is well known that (3.1) has for each $u(\cdot)$ a unique solution defined on $[t_0, t_f]$. This solution [8] is given by

$$x(t) = \Phi(t, t_0)\, x_0 + \int_{t_0}^{t} \Phi(t, \tau)\, B(\tau) u(\tau) d\tau, \tag{3.3}$$

where $\Phi(\cdot,\cdot)$ is an $n\times n$ matrix function of t and τ (the transition matrix corresponding to $A(\cdot)$) which satisfies the homogeneous ordinary differential equation

$$\begin{aligned} \frac{d}{dt}\Phi(t, \tau) &= A(t)\Phi(t, \tau), \\ \Phi(t_0, t_0) &= I. \end{aligned} \tag{3.4}$$

Let the control function $u^o(\cdot)$ generate the trajectory $x^o(\cdot)$, and suppose we perturb the control function $u^o(\cdot)$ by adding to it an arbitrary piecewise continuous function $\varepsilon\eta(\cdot)$, where ε is a small positive scalar parameter. Let the trajectory that results as a consequence of the control function $u^o(\cdot) + \varepsilon\eta(\cdot)$ be $x^o(\cdot) + \xi(\cdot;\varepsilon)$. We have from (3.3) that

$$x^o(t) + \xi(t;\varepsilon) = \Phi(t, t_0)x_0 + \int_{t_0}^{t} \Phi(t,\tau)B(\tau)\left[u^o(\tau) + \varepsilon\eta(\tau)\right] d\tau, \tag{3.5}$$

and it then follows that

$$\xi(t;\varepsilon) = \varepsilon \int_{t_0}^{t} \Phi(t,\tau)B(\tau)\eta(\tau)d\tau. \tag{3.6}$$

Upon defining

$$z(t;\eta(\cdot)) = \int_{t_0}^{t} \Phi(t,\tau)B(\tau)\eta(\tau)d\tau, \tag{3.7}$$

we see that

$$\xi(t;\varepsilon) = \varepsilon z(t;\eta(\cdot)), \tag{3.8}$$

so that by linearity perturbing $u^o(\cdot)$ by $\varepsilon\eta(\cdot)$ perturbs $x^o(\cdot)$ by a function of exactly the same form, viz., $\varepsilon z(t;\eta(\cdot))$.

Remark 3.2.2 *The notation used in this book can be compared to the variational notation like that used in* [11] *by noting (for example)*

$$u(\cdot) = u^o(\cdot) + \delta u(\cdot) \in \mathcal{U} \Rightarrow \|\delta u(\cdot)\| \leq \varepsilon_1, \delta u(\cdot) = \varepsilon\eta(\cdot). \tag{3.9}$$

3.2.2 Expansion Formula

The definition of a differentiable scalar function permits us to make the following Taylor expansion (see Appendix A.1.2) in the parameter ε:

$$\begin{aligned} &L(x^o(t) + \varepsilon z(t;\eta(\cdot)), u^o(t) + \varepsilon\eta(t), t) \\ &\quad = L(x^o(t), u^o(t), t) + \varepsilon L_x(x^o(t), u^o(t), t) z(t;\eta(\cdot)) \\ &\qquad + \varepsilon L_u(x^o(t), u^o(t), t)\eta(t) + \mathcal{O}(t;\varepsilon), \end{aligned} \tag{3.10}$$

where the function $\mathcal{O}(t;\varepsilon)$ is piecewise continuous in t and

$$\frac{\mathcal{O}(t;\varepsilon)}{\varepsilon} \to 0 \quad \text{as} \quad \varepsilon \to 0 \quad \text{for each } t.$$

3.2.3 Adjoining System Equation

We may adjoin (3.1) to the performance criterion (3.2) by means of a continuously differentiable n vector function of time $\lambda(\cdot)$, called in the classical literature the Lagrange multiplier (see Section 2.3.2), as follows:

$$\hat{J}(u(\cdot);\lambda(\cdot), x_0) = J(u(\cdot); x_0) + \int_{t_0}^{t_f} \lambda^T(t)\left[A(t)x(t) + B(t)u(t) - \dot{x}(t)\right] dt. \tag{3.11}$$

Note that when the differential constraint is satisfied,

$$\hat{J}\left(u(\cdot);\lambda(\cdot),x_0\right)=J\left(u(\cdot);x_0\right) \tag{3.12}$$

when (3.1) holds, so that nothing is gained or lost by this step. Because of the assumed differentiability of $\lambda(\cdot)$ with respect to t, we may integrate by parts in (3.11) to obtain

$$\begin{aligned}\hat{J}\left(u(\cdot);\lambda(\cdot),x_0\right) &= J\left(u(\cdot);x_0\right)+\int_{t_0}^{t_f}\left[\dot{\lambda}^T(t)x(t)+\lambda^T(t)A(t)x(t)+\lambda^T(t)B(t)u(t)\right]dt\\ &\quad+\lambda^T(t_0)x_0-\lambda^T(t_f)x_f. \end{aligned} \tag{3.13}$$

3.2.4 Expansion of $\widehat{J}$

We now evaluate the change in $\hat{J}$ (i.e., in J, since $\hat{J}=J$ when the dynamics are satisfied) brought about by changing u^o to $u^o(\cdot)+\varepsilon\eta(\cdot)$. Defining the change in the performance index as

$$\Delta\hat{J}\triangleq\hat{J}([u^o(\cdot)+\varepsilon\eta(\cdot)];\lambda(\cdot),x_0)-\hat{J}(u^o;\lambda(\cdot),x_0)$$

leads to the following expansion:

$$\begin{aligned}\Delta\hat{J} &= \int_{t_0}^{t_f}\Big[\varepsilon L_x(x^o(t),u^o(t),t)z(t;\eta(\cdot))+\varepsilon L_u(x^o(t),u^o(t),t)\eta(t)+\varepsilon\dot{\lambda}^T z(t,\eta(\cdot))\\ &\quad+\varepsilon\lambda^T(t)A(t)z(t;\eta(\cdot))+\varepsilon\lambda^T(t)B(t)\eta(t)+\mathcal{O}(t;\varepsilon)\Big]dt\\ &\quad-\varepsilon\lambda^T(t_f)z(t_f;\eta(\cdot))+\varepsilon\phi_x(x^o(t_f))z(t_f;\eta(\cdot))+\mathcal{O}(\varepsilon). \end{aligned} \tag{3.14}$$

Note that the coefficient of a first-order change in λ is zero, since it adjoins the system dynamics.

Now let us choose

$$\begin{aligned}-\dot{\lambda}^T(t) &= L_x(x^o(t),u^o(t),t)+\lambda^T(t)A(t),\\ \lambda^T(t_f) &= \phi_x(x^o(t_f)). \end{aligned} \tag{3.15}$$

This is a legitimate choice for $\lambda(\cdot)$ as (3.15) is a linear ordinary differential equation in λ with piecewise continuous coefficients, having a unique solution. The right-hand side of (3.14) then simplifies to

$$\begin{aligned}
\Delta \hat{J} &= \varepsilon \int_{t_0}^{t_f} \left[L_u(x^o(t), u^o(t), t)\eta(t) + \lambda^T(t)B(t)\eta(t) \right] dt \\
&\quad + \int_{t_0}^{t_f} \mathcal{O}(t;\varepsilon)dt + \mathcal{O}(\varepsilon) \qquad (3.16)\\
&= \varepsilon \delta \hat{J} + \int_{t_0}^{t_f} \mathcal{O}(t;\varepsilon)dt + \mathcal{O}(\varepsilon). \qquad (3.17)
\end{aligned}$$

3.2.5 Necessary Condition for Optimality

Let $u^o(\cdot)$ be the minimizing control. Therefore, $x^o(\cdot)$ is the resulting optimal path. As $\eta(\cdot)$ can be any piecewise continuous function on $[t_0, t_f]$ we may, because of our assumptions on L, λ, and B, set

$$\eta(t) = -\left[L_u(x^o(t), u^o(t), t) + \lambda^T(t)B(t) \right]^T. \qquad (3.18)$$

Substituting (3.18) into (3.16) yields

$$\Delta \hat{J} = -\varepsilon \int_{t_0}^{t_f} \left|\left| L_u(x^o(t), u^o(t), t) + \lambda^T(t)B(t) \right|\right|^2 dt + \int_{t_0}^{t_f} \mathcal{O}(t;\varepsilon)dt + \mathcal{O}(\varepsilon) \geq 0. \qquad (3.19)$$

For ε small and positive, the first term in (3.19) is dominant, so

$$-\varepsilon \int_{t_0}^{t_f} \left|\left| L_u(x^o(t), u^o(t), t) + \lambda^T(t)B(t) \right|\right|^2 dt \geq 0, \qquad (3.20)$$

where the variation is nonnegative because $u^o(\cdot)$ minimizes J. The left-hand side of (3.20) is negative if

$$L_u(x^o(t), u^o(t), t) + \lambda^T(t)B(t) \neq 0 \quad \forall t \in [t_0, t_f]. \qquad (3.21)$$

It follows, then, that a necessary condition for $u^o(\cdot)$ to minimize $\hat{J}$ and thus J is that

$$L_u(x^o(t), u^o(t), t) + \lambda^T(t)B(t) = 0 \qquad (3.22)$$

for all t in $[t_0, t_f]$. The multiplier $\lambda(\cdot)$ is obtained by (3.15) and satisfies the assumption that $\lambda(\cdot)$ is continuously differentiable.

3.2.6 Pontryagin's Necessary Condition for Weak Variations

The necessary condition for optimality of $u^o(\cdot)$ derived in Section 3.2.5 can be restated in terms of a particular Hamiltonian function to yield a *weak* version of Pontryagin's Principle for the optimal control problem formulated in Section 3.2. Let us define the Hamiltonian

$$H(x(t), u(t), \lambda(t), t) = L(x(t), u(t), t) + \lambda^T(t)[A(t)x(t) + B(t)u(t)]. \tag{3.23}$$

Then from (3.22) and (3.15) we have the following theorem.

Theorem 3.2.1 *Suppose that* $u^o(\cdot)$ *minimizes the performance criterion* (3.2). *Then, under Assumptions* 3.2.1–3.2.4, *the partial derivative of the Hamiltonian with respect to the control* $u(t)$ *is zero when evaluated at* $x^o(t)$, $u^o(t)$, *viz.,*

$$H_u(x^o(t), u^o(t), \lambda(t), t) = 0, \tag{3.24}$$

where

$$\begin{aligned} -\dot{\lambda}^T(t) &= L_x(x^o(t), u^o(t), t) + \lambda^T(t)A(t), \\ \lambda^T(t_f) &= \phi_x(x^o(t_f)). \end{aligned} \tag{3.25}$$

Note 3.2.2 *These necessary conditions have converted a functional minimization into a function minimization at each point in time.*

An Example Illustrating Pontryagin's Necessary Condition

Let us consider the system

$$\begin{aligned} \dot{x}_1(t) &= x_2(t), \quad x_1(t_0) = x_{10}, \\ \dot{x}_2(t) &= u(t), \quad x_2(t_0) = x_{20} \end{aligned} \tag{3.26}$$

and the performance functional

$$J(u(\cdot); x_0) = x_1(t_f) + \frac{1}{2}\int_{t_0}^{t_f} u^2(t)dt. \tag{3.27}$$

We guess that

$$\tilde{u}(t) = -(t_f - t) \tag{3.28}$$

is an optimizing control. In order to test whether this is a candidate for an optimal control, we check whether the necessary condition developed above is satisfied, although the necessary conditions can explicitly determine this control; i.e., see (3.32).

The Hamiltonian for this problem is given by

$$H(x, u, \lambda, t) = \frac{1}{2}u^2 + \lambda_1 x_2 + \lambda_2 u \tag{3.29}$$

so that

$$\begin{aligned} -\dot{\lambda}_1(t) &= H_{x_1} = 0, & \lambda_1(t_f) &= 1, \\ -\dot{\lambda}_2(t) &= H_{x_2} = \lambda_1(t), & \lambda_2(t_f) &= 0. \end{aligned} \tag{3.30}$$

Hence we obtain

$$\lambda_1(t) = 1, \quad \lambda_2(t) = (t_f - t). \tag{3.31}$$

From (3.29) we have that

$$H_u(x, u, \lambda, t) = u + \lambda_2, \tag{3.32}$$

and it follows from Equations (3.28), (3.31), and (3.32) that

$$H_u(\tilde{x}(t), \tilde{u}(t), \tilde{\lambda}(t), t) = 0. \tag{3.33}$$

Note that we have not shown that $\tilde{u}(\cdot)$ minimizes J but only that it satisfies a condition that an optimizing control must satisfy, i.e., that it satisfies a necessary condition for optimality.

Using the Lagrange multiplier method we now prove directly that $\tilde{u}(\cdot)$ is indeed minimizing. We adjoin (3.26) to J using $\lambda(\cdot)$ and integrate by parts to obtain

$$\begin{aligned}
&\hat{J}(u(\cdot); \lambda(\cdot), x_0) \\
&= \int_{t_0}^{t_f} \left[\frac{1}{2} u^2(t) + \lambda_1(t) x_2(t) + \lambda_2(t) u(t) + \dot{\lambda}_1(t) x_1(t) + \dot{\lambda}_2(t) x_2(t) \right] dt \\
&\quad + x_1(t_f) + \lambda_1(t_0) x_{10} + \lambda_2(t_0) x_{20} - \lambda_1(t_f) x_1(t_f) - \lambda_2(t_f) x_2(t_f). \qquad (3.34)
\end{aligned}$$

Using (3.30) and (3.31) to specify $\lambda(\cdot)$ in (3.34) yields

$$\begin{aligned}
\hat{J}(u(\cdot); \lambda(\cdot), x_0) &= \int_{t_0}^{t_f} \left[\frac{1}{2} u^2(t) + (t_f - t) u(t) \right] dt \\
&\quad + \lambda_1(t_0) x_{10} + \lambda_2(t_0) x_{20}. \qquad (3.35)
\end{aligned}$$

Now the only term that depends upon $u(\cdot)$ is the integral, which we can write as

$$\frac{1}{2} \int_{t_0}^{t_f} \left[(u + t_f - t)^2 - (t_f - t)^2 \right] dt, \qquad (3.36)$$

and this clearly takes on its minimum value when we set

$$u(t) = \tilde{u}(t) = -(t_f - t), \qquad (3.37)$$

and the minimum value of $\hat{J}(u(\cdot); \lambda(\cdot), x_0)$ is then

$$\begin{aligned}
\min_u \hat{J}(u(\cdot); \lambda(\cdot), x_0) &= -\frac{1}{2} \int_{t_0}^{t_f} (t_f - t)^2 dt + \lambda_1(0) x_{10} + \lambda_2(0) x_{20} \\
&= -\frac{1}{6} (t_f - t_0)^3 + x_{10} + (t_f - t_0) x_{20}. \qquad (3.38)
\end{aligned}$$

Problems

1. In the example above, assume $\lambda(t) = P(t)x(t)$. Find a differential equation for P and a feedback control law in the form $u^o(t) = \Lambda(t)x(t)$, i.e., determine $\Lambda(t)$.

2. Consider the cost criterion

$$J = \phi_1(x(t_f)) + \phi_2(x(t_1)) + \int_{t_0}^{t_f} L(x, u, t)dt \tag{3.39}$$

subject to

$$\dot{x} = Ax + Bu, \quad x_0 = x(t_0) \text{ given}, \tag{3.40}$$

and $t_0 < t_1 < t_f$ with t_1 and t_f fixed. Determine the Pontryagin necessary conditions for this problem.

3.3 Nonlinear Dynamic System

We here extend the results thus far obtained to the case where the dynamic system is nonlinear, described by

$$\begin{aligned} \dot{x}(t) &= f(x(t), u(t), t), \\ x(t_0) &= x_0, \end{aligned} \tag{3.41}$$

where $f(\cdot, \cdot, \cdot)$ is an n vector function of its arguments. Our first task is clearly to investigate the behavior of this nonlinear ordinary differential equation in a neighborhood of a trajectory-control pair $(x^o(\cdot),\ u^o(\cdot))$ in order to obtain an expression analogous to (3.8). We make the following assumptions.

Assumption 3.3.1 *The n vector function $f(\cdot, \cdot, \cdot)$ is once continuously differentiable in x and u and continuous in t on the interval $[t_0, t_f]$.*

Assumption 3.3.2 *Equation (3.41) has a unique solution $x(\cdot)$ defined on $[t_0, t_f]$ for each piecewise continuous control function $u(\cdot)$.*

Actually, it is necessary only to assume existence on $[t_0, t_f]$, as uniqueness of solutions follows from Assumption 3.3.1. Note that it is necessary in the case of nonlinear dynamic systems to make an assumption about existence of a solution on the interval $[t_0, t_f]$. For example, the quadratic equation

$$\dot{x}(t) = x(t)^2, \quad x(t_0) = x_0 > 0, \tag{3.42}$$

has the unique solution

$$x(t) = \frac{x_0}{1 - x_0(t - t_0)}, \tag{3.43}$$

which ceases to exist at $t = \frac{1}{x_0} + t_0$. This implies that (3.42) does not have a solution defined on $[t_0, t_f]$ if $t_f \geq \frac{1}{x_0} + t_0$ (a finite escape time).

3.3.1 Perturbations in the Control and State from the Optimal Path

Let the control function $u^o(\cdot)$ generate the unique trajectory $x^o(\cdot)$, and suppose that we perturb this control function by adding to it a piecewise continuous function $\varepsilon\eta(\cdot)$, where ε is a positive scalar parameter. Let the trajectory that results as a consequence of the control function $u^o(\cdot) + \varepsilon\eta(\cdot)$ be $x^o(\cdot) + \xi(\cdot;\varepsilon)$, where $\xi(t_0;\varepsilon) = 0$. From (3.41) we have that

$$\dot{x}^o(t) + \dot{\xi}(t;\varepsilon) = f([x^o(t) + \xi(t;\varepsilon)], [u^o(t) + \varepsilon\eta(t)], t), \tag{3.44}$$

so that

$$\dot{\xi}(t;\varepsilon) = f([x^o(t) + \xi(t;\varepsilon)], [u^o(t) + \varepsilon\eta(t)], t) - f(x^o(t), u^o(t), t). \tag{3.45}$$

Now because $f(\cdot,\cdot,\cdot)$ satisfies Assumption 3.3.1, it is well known in differential equation theory [6] that $\xi(\cdot;\varepsilon)$ is once continuously differentiable with respect to ε (Bliss's

Theorem). Consequently we can write

$$\xi(t;\varepsilon) = \varepsilon z(t;\eta(\cdot)) + \mathcal{O}(t;\varepsilon), \tag{3.46}$$

where, from (3.45), the partial derivative of $\xi(\cdot;\varepsilon)$ with respect to ε, $\xi_\varepsilon(\cdot)$ is propagated as

$$\dot{\xi}_\varepsilon(t) = f_\xi(x^o(t), u^o(t), t)\xi_\varepsilon(t) + f_u(x^o(t), u^o(t), t)\eta(t), \tag{3.47}$$

where $\xi_\varepsilon(\cdot) \triangleq z(t;\eta(\cdot))$. Having established this intermediate result, which is analogous to (3.8), we expand $f(\cdot,\cdot,t)$ as follows:

$$\begin{aligned} f([x^o(t) + \xi(t;\varepsilon)], [u^o(t) + \varepsilon\eta(t)], t) &= f(x^o(t), u^o(t), t) \\ + \quad \varepsilon f_x(x^o(t), u^o(t), t) z(t;\eta(\cdot)) &+ \varepsilon f_u(x^o(t), u^o(t), t)\eta(t) + \mathcal{O}(t;\varepsilon). \end{aligned} \tag{3.48}$$

Adjoin (3.41) to the performance criterion (3.2) as follows:

$$\hat{J}(u(\cdot);\lambda(\cdot), x_0) = J(u(\cdot); x_0) + \int_{t_0}^{t_f} \lambda^T(t)\left[f(x(t), u(t), t) - \dot{x}(t)\right] dt, \tag{3.49}$$

where $\lambda(t)$ is an as yet undetermined, continuously differentiable, n vector function of time on $[t_0, t_f]$. Integrating by parts we obtain

$$\begin{aligned} \hat{J}(u(\cdot);\lambda(\cdot), x_0) &= J(u(\cdot); x_0) + \int_{t_0}^{t_f} \left[\dot{\lambda}^T(t)x(t) + \lambda^T(t) f(x(t), u(t), t)\right] dt \\ &\quad + \lambda^T(t_0)x_0 - \lambda^T(t_f)x(t_f). \end{aligned} \tag{3.50}$$

Evaluate the change in $\hat{J}$ (i.e., in J) brought about by changing $u^o(\cdot)$ to $u^o(\cdot) + \varepsilon\eta(\cdot)$:

$$\begin{aligned} \Delta\hat{J} &\triangleq \hat{J}(u^o(\cdot) + \varepsilon\eta(\cdot);\lambda(\cdot), x_0) - \hat{J}(u^o(\cdot);\lambda(\cdot), x_0) \\ &= \int_{t_0}^{t_f} \Big[\varepsilon L_x(x^o(t), u^o(t), t) z(t;\eta(\cdot)) + \varepsilon\lambda^T(t) f_x(x^o(t), u^o(t), t) z(t;\eta(\cdot)) \\ &\qquad + \varepsilon L_u(x^o(t), u^o(t), t)\eta(t) + \varepsilon\lambda^T(t) f_u(x^o(t), u^o(t), t)\eta(t) \\ &\qquad + \varepsilon\dot{\lambda}^T(t) z(t;\eta(\cdot)) + \mathcal{O}(t;\varepsilon)\Big] dt \\ &\quad - \varepsilon\lambda^T(t_f) z(t_f;\eta(\cdot)) + \varepsilon\phi_x(x^o(t_f)) z(t_f;\eta(\cdot)) + \mathcal{O}(\varepsilon). \end{aligned} \tag{3.51}$$

Define the variational Hamiltonian as

$$H\left(x(t), u(t), \lambda(t), t\right) = L\left(x(t), u(t), t\right) + \lambda^T(t) f\left(x(t), u(t), t\right), \tag{3.52}$$

and set

$$\begin{aligned} -\dot{\lambda}^T(t) &= H_x\left(x^o(t), u^o(t), \lambda(t), t\right), \\ \lambda^T(t_f) &= \phi_x(x(t_f)), \end{aligned} \tag{3.53}$$

which, as with (3.15), is legitimate. The right-hand side of (3.51) then becomes

$$\Delta J = \varepsilon \int_{t_0}^{t_f} \left[H_u(x^o(t), u^o(t), \lambda(t), t)\, \eta(t)\right] dt + \int_{t_0}^{t_f} \mathcal{O}(t;\varepsilon) dt + \mathcal{O}(\varepsilon), \tag{3.54}$$

which is identical in form to (3.16). The same reasoning used in Section 3.2.5 then yields the following weak version of Pontryagin's Principle.

3.3.2 Pontryagin's Weak Necessary Condition

Theorem 3.3.1 *Suppose $u^o(\cdot)$ minimizes the performance criterion* (3.2) *subject to the dynamic system* (3.41) *and that $H(x(t), u(t), \lambda(t), t)$ is defined according to* (3.52). *Then, under Assumptions* 3.2.2–3.2.4 *and* 3.3.1–3.3.2, *the partial derivative of the Hamiltonian with respect to the control $u(t)$ is zero when evaluated at $x^o(t)$, $u^o(t)$, viz.,*

$$H_u\left(x^o(t), u^o(t), \lambda(t), t\right) = 0, \tag{3.55}$$

where

$$-\dot{\lambda}^T(t) = H_x\left(x^o(t), u^o(t), \lambda(t), t\right), \tag{3.56}$$

$$\lambda^T(t_f) = \phi_x(x(t_f)). \tag{3.57}$$

Clearly this Theorem is exactly the same as Theorem 3.2.1 with $A(t)x(t) + B(t)u(t)$ replaced by $f(x(t), u(t), t)$. However, it is necessary to invoke Bliss's Theorem on differentiability with respect to a parameter of the solution of nonlinear ordinary differential equations; this was not required in Section 3.2.1, where the linearity of (3.1) sufficed.

Remark 3.3.1 *If H as defined in (3.52) is not an explicit function of t, i.e., $H(x(t), u(t), \lambda(t), t) = H(x(t), u(t), \lambda(t))$ and $u(\cdot) \in \mathcal{U}_c$, the class of continuously differentiable functions, then $H(x(t), u(t), \lambda(t))$ is a constant of the motion along the optimal path. This is easily shown by noting that $\dot{H}(x(t), u(t), \lambda(t)) = 0$ by using (3.41), (3.55), and (3.56).*

3.3.3 Maximum Horizontal Distance: A Variation of the Brachistochrone Problem

We return to the brachistochrone example of Section 2.1. Here, we take a slightly modified form, where we maximize the horizontal distance $r(t_f)$ traveled in a fixed time t_f. In this formulation, we have the dynamics

$$\dot{r} = v\cos\theta, \qquad r(0) = 0, \tag{3.58}$$

$$\dot{z} = v\sin\theta, \qquad z(0) = 0, \tag{3.59}$$

$$\dot{v} = g\sin\theta, \qquad v(0) = 0, \tag{3.60}$$

and the cost criterion is

$$J = \phi(x(t_f)) = -r(t_f), \tag{3.61}$$

where v is the velocity and θ is the control variable.

Necessary Conditions

We first note that the value of z appears in neither the dynamics nor the performance index and may therefore be ignored. We thus reduce the state space to

$$x = \begin{bmatrix} r \\ v \end{bmatrix}. \tag{3.62}$$

The dynamics can be written

$$\dot{x} = f(x, \theta) = \begin{bmatrix} v\cos\theta \\ g\sin\theta \end{bmatrix}. \tag{3.63}$$

The variational Hamiltonian for this problem is

$$H(x, u, t, \lambda) = \lambda^T f = \lambda_r v\cos\theta + \lambda_v g\sin\theta, \tag{3.64}$$

where we have taken the elements of the Lagrange multiplier vector to be λ_r and λ_v. Since H is invariant with respect to time, $H = C$, where C is a constant (see Remark 3.3.1). The necessary conditions from Theorem 3.3.1 are

$$\dot{\lambda}_r = 0, \qquad \lambda_r(t_f) = -1, \tag{3.65}$$

$$\dot{\lambda}_v = -\lambda_r\cos\theta, \qquad \lambda_v(t_f) = 0, \tag{3.66}$$

and

$$\frac{\partial H}{\partial \theta} = -\lambda_r v\sin\theta + \lambda_v g\cos\theta = 0. \tag{3.67}$$

From (3.65), we have

$$\lambda_r(t) \equiv -1, \quad t \in [0, t_f], \tag{3.68}$$

and from (3.67),

$$\tan\theta(t) = \frac{\lambda_v(t)g}{\lambda_r v(t)} = -\frac{\lambda_v(t)g}{v(t)}. \tag{3.69}$$

Note that $v(0) = 0$ implies that $\theta(0) = \pi/2$, and $\lambda_v(t_f) = 0 \implies \theta(t_f) = 0$. These are characteristics of the classic solution to the problem.

This problem can be solved in closed form. From (3.69) and $H = C$, (3.64) reduces to

$$\cos\theta = -\frac{v}{C}. \tag{3.70}$$

Using (3.69) and (3.70) together gives

$$\sin\theta = \frac{\lambda_v g}{C}. \tag{3.71}$$

Introducing (3.70) into (3.58) gives

$$\dot{r} = -\frac{v^2}{C} \tag{3.72}$$

and into (3.60) and (3.66) results in an oscillator as

$$\begin{bmatrix} \dot{v} \\ \dot{\lambda}_v \end{bmatrix} = \begin{bmatrix} 0 & g^2/C \\ -1/C & 0 \end{bmatrix} \begin{bmatrix} v \\ \lambda_v \end{bmatrix}, \quad \begin{bmatrix} v(0) \\ \lambda_v(t_f) \end{bmatrix} = \begin{bmatrix} 0 \\ 0 \end{bmatrix}. \tag{3.73}$$

The solution to (3.73) is

$$\lambda_v(t) = a_1 \sin\frac{g}{C}t + a_2\cos\frac{g}{C}t, \tag{3.74}$$

$$-C\dot{\lambda}_v = v(t) = -a_1 g\cos\frac{g}{C}t + a_2 g\sin\frac{g}{C}t. \tag{3.75}$$

At $t = 0$, $v(0) = 0$ and, therefore, $a_1 = 0$. At $t = t_f$, since $\theta(t_f) = 0$, from (3.70) $C = -v(t_f)$. Since $v(t_f)$ is positive and thus C is negative, $\lambda_v(t_f) = 0 = a_2\cos(\frac{gt_f}{C}) \Rightarrow \frac{gt_f}{C} = -\frac{\pi}{2} \Rightarrow \sin\frac{gt_f}{C} = -1$. Therefore, $v(t_f) = -ga_2 \neq 0$. To determine C, note that at t_f, $\theta(t_f) = 0$. Then

$$C = -v(t_f) = \frac{-2gt_f}{\pi} = ga_2 \Rightarrow a_2 = -\frac{2t_f}{\pi}. \tag{3.76}$$

Since $v(t) = \frac{2gt_f}{\pi}\sin\frac{gt}{C}$, then

$$\dot{r} = v\cos\theta = -\frac{v^2}{C} = \frac{2gt_f}{\pi}\sin^2\frac{\pi t}{2t_f}. \tag{3.77}$$

This integrates to

$$r(t) = \frac{2gt_f}{\pi} \int_0^{t_f} \sin^2 \frac{\pi t}{2t_f} dt = \frac{g\tau^2}{4} \left[\frac{2t}{\tau} - \sin \frac{2t}{\tau} \right], \tag{3.78}$$

where $\tau = \frac{2t_f}{\pi}$. The maximum horizontal distance is $r(t_f) = \frac{gt_f^2}{\pi}$. In the vertical direction,

$$\dot{z} = v \sin \theta = \frac{v\dot{v}}{g} = g\tau \sin \frac{t}{\tau} \cos \frac{t}{\tau} = \frac{g\tau}{2} \sin \frac{2t}{\tau}. \tag{3.79}$$

This integrates to

$$z(t) = g \left[1 - \cos \frac{2t}{\tau} \right], \tag{3.80}$$

where the value at t_f is $z(t_f) = 2g$. Note that (3.78) and (3.80) are the equations for a cycloid.

3.3.4 Two-Point Boundary-Value Problem

In Section 3.2.6 we verified that a certain control function, for a specific example, satisfied the conditions of Theorem 3.2.1. This was possible without resorting to numerical (computer) methods because of the simplicity of the example.

For a general nonlinear dynamic system this "verification" is also possible. First, one integrates (3.41), numerically if necessary, with $u(\cdot) = \tilde{u}(\cdot)$, where $\tilde{u}(\cdot)$ is the control function to be tested, to obtain $\tilde{x}(\cdot)$. One then integrates (3.56) backward in time along the path $\tilde{x}(\cdot)$, $\tilde{u}(\cdot)$ from the "initial" value $\lambda^T(t_f) = \phi_x(\tilde{x}(t_f))$. Having done this, one then has in hand $\tilde{x}(\cdot)$, $\tilde{u}(\cdot)$, and $\lambda(\cdot)$ so that (3.55) can be tested.

One might also wish to generate (construct) a control function $\tilde{u}(\cdot)$ which satisfies Pontryagin's Principle. This is more difficult, since the initial condition for (3.41) is known at $t = t_0$, whereas only the final condition for (3.56) is known at $t = t_f$, i.e., we have a so-called two-point boundary-value problem: (3.41) and (3.56) cannot be integrated simultaneously from $t = t_0$ or from $t = t_f$. Usually one has to resort to numerical techniques to solve the two-point boundary-value problem. The so-called linear quadratic optimal control problem is one class of problems in which the two-point boundary-value problem can be converted into an ordinary initial value problem; this class of problems is discussed in Chapter 5.

A numerical optimization method is presented which may converge to a local optimal path. This method, called steepest descent, iterates on H_u until the optimality condition $H_u = 0$ is approximately satisfied. The difficulty with this approach is that as H_u becomes small, the converges rate becomes slow. For comparison, a second numerical optimization method called the shooting method is described in Section 5.4.7; it satisfies the optimality condition explicitly on each iteration but requires converging to the boundary conditions. In contrast to the steepest descent method, the shooting method converges very fast in the vicinity of the the optimal path. However, if the initial choice of the boundary conditions is quite far from that of the optimal path, convergence can be very slow. Steepest descent usually converges initially well from the guessed control sequence. Another difficulty with the shooting method is that it may try to converge to an extremal trajectory that satisfies the first-order necessary conditions but is not locally minimizing (see Chapter 5). The steepest descent method does not converge to an extremal trajectory that satisfies the first-order necessary conditions but is not locally minimizing.

Solving the Two-Point Boundary-Value Problem via the Steepest Descent Method

The following is used in the procedure of the following steepest descent algorithm.

Note 3.3.1 *The relationship between a first-order perturbation in the cost and a perturbation in the control is*

$$\begin{aligned}\delta J &= \phi_x(x^i(t_f))\delta x(t_f) + \int_{t_0}^{t_f} (L_x(x^i(t), u^i(t), t)\delta x(t) + L_u(x^i(t), u^i(t), t)\delta u)dt \\ &= \int_{t_0}^{t_f} (L_u(x^i(t), u^i(t), t) + \lambda^{iT}(t) f_u(x^i(t), u^i(t), t))\delta u dt, \end{aligned} \tag{3.81}$$

where the perturbed cost (3.81) *is obtained from the first-order perturbation* δx *generated by the linearized differential equation*

$$\delta \dot{x}(t) = f_x(x^i(t), u^i(t), t)\delta x(t) + f_u(x^i(t), u^i(t), t)\delta u(t), \tag{3.82}$$

adding the identically zero term $\int_{t_0}^{t_f} \frac{d}{dt}(\lambda^{iT}\delta x)dt - \phi_x(x^i(t_f))\delta x(t_f)$ *to the perturbed cost, and by assuming that the initial condition is given.*

The steps in the steepest descent algorithm are given in the following procedure.

1. Choose the nominal control, $u^i \in \mathcal{U}$, at iteration $i = 1$.

2. Generate the nominal state path, $x^i(\cdot)$, i.e., integrate forward from t_0 to t_f.

3. Integrate the adjoint λ^i equation backward along the nominal path

$$\lambda^i(t_f) = \phi_x^T(x^i(t_f)), \tag{3.83}$$

$$\dot{\lambda}^i(t) = -L_x(x^i(t), u^i(t), t)^T - f_x(x^i(t), u^i(t), t)^T \lambda^i(t). \tag{3.84}$$

4. From Note 3.3.1 let

$$H_u(x^i(t), u^i(t), t) = L_u(x^i(t), u^i(t), t) + \lambda^{iT}(t) f_u(x^i(t), u^i(t), \lambda^i, t). \tag{3.85}$$

   Form the control variation $\delta u(t) = -\epsilon H_u^T(x^i(t), u^i(t), t)$ to determine the control for the $(i+1)$th iteration as

$$u^{i+1}(t) = u^i(t) - \epsilon H_u^T(x^i(t), u^i(t), t) \tag{3.86}$$

for some choice of $\epsilon > 0$, which preserves the assumed linearity. The perturbed cost criterion (3.81) is approximately

$$\delta J = -\epsilon \int_{t_0}^{t_f} \|H_u(x^i(t), u^i(t), t)\|^2 dt. \tag{3.87}$$

5. If $\|H_u(x^i(t), u^i(t), t)\|$ is *sufficiently small* over the path, then stop. If not, go to 2.

Problems

Solve for the control $u(\cdot) \in \mathcal{U}$ that minimizes

$$J = \frac{1}{2}\left[x(10)^2 + \int_0^{10} (x^2 + 2bxu + u^2)dt\right]$$

subject to the scalar state equation

$$\dot{x} = x + u, \quad x(0) = x_0 = 1$$

1. analytically with $b = 0, -1, -10$;

2. numerically by steepest descent by using the algorithm in Section 3.3.4.

3.4 Strong Variations and the Strong Form of the Pontryagin Minimum Principle

Theorem 3.3.1 is referred to as a weak Pontryagin Principle because it states only that

$$H_u\left(x^o(t), u^o(t), \lambda(t), t\right) = 0. \tag{3.88}$$

It turns out, however, that a stronger statement is possible. That is, if $u^o(\cdot)$ minimizes J, then

$$H\left(x^o(t), u^o(t), \lambda(t), t\right) \le H\left(x^o(t), u(t), \lambda(t), t\right) \quad \forall t \in [t_0, t_f]. \tag{3.89}$$

Clearly (3.89) implies (3.88) but (3.89) is stronger as it states that $u^o(t)$ minimizes with respect to $u(t)$ the function $H(x(t), u(t), \lambda(t), t)$.

In the following derivation, the assumption of continuous differentiability with respect to u can be relaxed. Therefore, Assumptions 3.2.3 and 3.3.1 are replaced by the following.

Assumption 3.4.1 *The scalar function $L(\cdot,\cdot,\cdot)$ and the n vector function $f(\cdot,\cdot,\cdot)$ is once continuously differentiable in x and continuous in u and t where $t \in [t_0, t_f]$.*

In order to prove this strong form of Pontryagin's Principle one introduces the notion of strong perturbations (variations). Here a perturbation $\eta(\cdot)$ is made to $u^o(\cdot)$ which may be large in magnitude but which is nonzero only over a small time interval ε. In other words, instead of introducing a perturbation $\varepsilon\eta(\cdot)$ which can be made small by making ε small, we introduce a not necessarily small continuous perturbation $\eta(t)$, $\bar{t} \le t \le \bar{t}+\varepsilon$ and set $\eta(\cdot) = 0$ on the intervals $[t_0, \bar{t}\,)$, $(\bar{t}+\varepsilon, t_f]$. As we will now show, this type of perturbation still results in a small change in $x(\cdot)$.

Since $\eta(\cdot) = 0$ on $[t_0, \bar{t}\,)$, we have $x(t) = x^o(t)$ on $[t_0, \bar{t}\,]$ and $\xi(t;\eta(\cdot)) \triangleq x(t) - x^o(t)$ is zero on this interval. On the interval $[\bar{t}, \bar{t}+\varepsilon]$ we have

$$\dot{\xi}\,(t;\eta(\cdot)) = f\,(x^o(t) + \xi(t;\eta(\cdot)), u^o(t) + \eta(t), t) - f\,(x^o(t), u^o(t), t) \tag{3.90}$$

and

$$\xi\left(\bar{t};\eta(\cdot)\right) = 0,$$

so that

$$\begin{aligned} \xi\,(t;\eta(\cdot)) \quad &= \quad \int_{\bar{t}}^{t} [f\,(x^o(\tau) + \xi(t;\eta(\cdot)), u^o(\tau) + \eta(\tau), \tau) \\ &\quad - f\,(x^o(\tau), u^o(\tau), \tau)]\, d\tau \end{aligned} \tag{3.91}$$

for $\bar{t} \le t \le \bar{t}+\varepsilon$.

Clearly, the piecewise differentiability of $\xi(\cdot;\eta(\cdot))$ with respect to t allows us to write

$$\xi\,(t;\eta(\cdot)) = (t-\bar{t})\dot{\xi}\left(\bar{t};\eta(\cdot)\right) + \mathcal{O}(t-\bar{t}) \tag{3.92}$$

and

$$\xi\left(\bar{t}+\varepsilon;\eta(\cdot)\right)=\varepsilon\dot{\xi}\left(\bar{t};\eta(\cdot)\right)+\mathcal{O}(\varepsilon). \tag{3.93}$$

It then follows along the lines indicated in Section 3.3 that for $t\in[\bar{t}+\varepsilon,t_f]$

$$\xi\left(t;\eta(\cdot)\right)=\varepsilon z\left(t;\eta(\cdot)\right)+\mathcal{O}(t;\varepsilon). \tag{3.94}$$

Because of its differentiability, we can expand L for $\bar{t}\le t\le\bar{t}+\varepsilon$ as

$$\begin{aligned}
&L\left(\left[x^o(t)+(t-\bar{t})\dot{\xi}\left(\bar{t};\eta(\cdot)\right)+\mathcal{O}(t-\bar{t})\right],\left[u^o(t)+\eta(t)\right],t\right)\\
&\quad=L\left(x^o(t),\left[u^o(t)+\eta(t)\right],t\right)+(t-\bar{t})L_x\left(x^o(t),\left[u^o(t)+\eta(t)\right],t\right)\dot{\xi}\left(\bar{t};\eta(\cdot)\right)\\
&\qquad+\mathcal{O}(t-\bar{t}),
\end{aligned} \tag{3.95}$$

and for $\bar{t}+\varepsilon<t\le t_f$

$$\begin{aligned}
&L\left(\left[x^o(t)+\varepsilon z\left(t;\eta(\cdot)\right)+\mathcal{O}(t;\varepsilon)\right],u^o(t),t\right)\\
&\quad=L\left(x^o(t),u^o(t),t\right)+\varepsilon L_x\left(x^o(t),u^o(t),t\right)z\left(t;\eta(\cdot)\right)+\mathcal{O}(t;\varepsilon).
\end{aligned} \tag{3.96}$$

Similarly, for $\bar{t}\le t\le\bar{t}+\varepsilon$

$$\begin{aligned}
&f\left(\left[x^o(t)+(t-\bar{t})\dot{\xi}\left(\bar{t};\eta(\cdot)\right)+\mathcal{O}(t-\bar{t})\right],\left[u^o(t)+\eta(t)\right],t\right)\\
&\quad=f\left(x^o(t),\left[u^o(t)+\eta(t)\right],t\right)+(t-\bar{t})f_x\left(x^o(t),\left[u^o(t)+\eta(t)\right],t\right)\dot{\xi}\left(\bar{t};\eta(\cdot)\right)\\
&\qquad+\mathcal{O}(t-\bar{t}),
\end{aligned} \tag{3.97}$$

and for $\bar{t}+\varepsilon<t\le t_f$

$$\begin{aligned}
&f\left(\left[x^o(t)+\varepsilon z\left(t;\eta(\cdot)\right)+\mathcal{O}(t;\varepsilon)\right],u^o(t),t\right)\\
&\quad=f\left(x^o(t),u^o(t),t\right)+\varepsilon f_x\left(x^o(t),u^o(t),t\right)z\left(t;\eta(\cdot)\right)+\mathcal{O}(t;\varepsilon).
\end{aligned} \tag{3.98}$$

Using (3.50) we evaluate the change in $\hat{J}$ (i.e., in J) brought about by the strong variation $u^o(\cdot)$ to $u^o(\cdot)+\eta(\cdot)$ over $t\in[\bar{t},\bar{t}+\varepsilon]$, whereas $\eta(\cdot)=0$ for $t\in[t_0,\bar{t})$ and $t\in(\bar{t}+\varepsilon,t_f]$:

$$
\begin{aligned}
\Delta\hat{J}(\cdot,\cdot,\cdot) &\triangleq \hat{J}(u^o(\cdot)+\eta(\cdot);\lambda(\cdot),x_0)-\hat{J}(u^o(\cdot);\lambda(\cdot),x_0)\\
&= \int_{\bar{t}}^{\bar{t}+\varepsilon}\Big[L\left(x^o(t),[u^o(t)+\eta(t)],t\right)-L\left(x^o(t),u^o(t),t\right)\\
&\qquad +(t-\bar{t})L_x\left(x^o(t),[u^o(t)+\eta(t)],t\right)\dot{\xi}\left(\bar{t};\eta(\cdot)\right)\\
&\qquad +\lambda^T(t)\left[f\left(x^o(t),[u^o(t)+\eta(t)],t\right)-f\left(x^o(t),u^o(t),t\right)\right]\\
&\qquad +(t-\bar{t})\lambda^T(t)f_x\left(x^o(t),[u^o(t)+\eta(t)],t\right)\dot{\xi}\left(\bar{t};\eta(\cdot)\right)\\
&\qquad +(t-\bar{t})\dot{\lambda}^T(t)\dot{\xi}\left(\bar{t};\eta(\cdot)\right)+\mathcal{O}(t-\bar{t})\Big]\,dt\\
&\quad +\int_{\bar{t}+\varepsilon}^{t_f}\Big[\varepsilon L_x\left(x^o(t),u^o(t),t\right)z(t;\eta(\cdot))+\varepsilon\lambda^T(t)f_x\left(x^o(t),u^o(t),t\right)z(t;\eta(\cdot))\\
&\qquad +\varepsilon\dot{\lambda}^T(t)z(t;\eta(\cdot))+\mathcal{O}(t;\varepsilon)\Big]\,dt\\
&\quad +\varepsilon\lambda^T(t_f)z(t_f;\eta(\cdot))+\varepsilon\phi_x(x^o(t_f))z(t_f;\eta(\cdot))+\mathcal{O}(\varepsilon). \qquad (3.99)
\end{aligned}
$$

By using the definition of H from (3.52) and applying (3.53) over the interval $(\bar{t}+\varepsilon, t_f]$ in (3.99), we obtain

$$
\begin{aligned}
\Delta\hat{J} &= \int_{\bar{t}}^{\bar{t}+\varepsilon}\Big\{H\left(x^o(t),[u^o(t)+\eta(t)],\lambda(t),t\right)-H\left(x^o(t),u^o(t),\lambda(t),t\right)\\
&\qquad +\Big[(t-\bar{t})H_x\left(x^o(t),[u^o(t)+\eta(t)],\lambda(t),t\right)+(t-\bar{t})\dot{\lambda}^T(t)\Big]\dot{\xi}\left(\bar{t};\eta(\cdot)\right)\\
&\qquad +\mathcal{O}(t-\bar{t})\Big\}\,dt+\int_{\bar{t}+\varepsilon}^{t_f}\mathcal{O}(t;\varepsilon)dt+\mathcal{O}(\varepsilon). \qquad (3.100)
\end{aligned}
$$

The first integral in (3.100) can be expanded in terms of ε around the point $\bar{t}+\varepsilon$ to yield

$$
\begin{aligned}
\Delta\hat{J} &= \varepsilon\Big[H\left(x^o(\bar{t}+\varepsilon),[u^o(\bar{t}+\varepsilon)+\eta(\bar{t}+\varepsilon)],\lambda(\bar{t}+\varepsilon),(\bar{t}+\varepsilon)\right)\\
&\qquad -H\left(x^o(\bar{t}+\varepsilon),u^o(\bar{t}+\varepsilon),\lambda(\bar{t}+\varepsilon),(\bar{t}+\varepsilon)\right)\Big]\\
&\quad +\varepsilon^2\Big[H_x\left(x^o(\bar{t}+\varepsilon),[u^o(\bar{t}+\varepsilon)+\eta(\bar{t}+\varepsilon)],\lambda(\bar{t}+\varepsilon),(\bar{t}+\varepsilon)\right)\\
&\qquad +\dot{\lambda}^T(\bar{t}+\varepsilon)\Big]\dot{\xi}\left(\bar{t};\eta(\cdot)\right)+\varepsilon\mathcal{O}(\varepsilon)+\int_{\bar{t}+\varepsilon}^{t_f}\mathcal{O}(t;\varepsilon)dt+\mathcal{O}(\varepsilon). \qquad (3.101)
\end{aligned}
$$

For small ε, the dominant term in (3.101) is the first one, and as (3.101) must be nonnegative for the left-hand side of (3.99) to be nonnegative, we have the necessary condition

$$\begin{aligned} &H(x^o(\bar{t}+\varepsilon), u^o(\bar{t}+\varepsilon)+\eta(\bar{t}+\varepsilon), \lambda(\bar{t}+\varepsilon), \bar{t}+\varepsilon) \\ &\geq H(x^o(\bar{t}+\varepsilon), u^o(\bar{t}+\varepsilon), \lambda(\bar{t}+\varepsilon), \bar{t}+\varepsilon). \end{aligned} \tag{3.102}$$

As $\bar{t}$ and ε are arbitrary and as the value of $\eta(\bar{t}+\varepsilon)$ is arbitrary, we conclude that $H(x^o(t), u(t), \lambda(t), t)$ is minimized with respect to $u(t)$ at $u(t) = u^o(t)$. We have thus proved the following theorem.

Theorem 3.4.1 *Suppose $u^o(\cdot)$ minimizes the performance criteria* (3.2) *subject to the dynamic system* (3.41) *and that $H(x, u, \lambda, t)$ is defined according to* (3.52). *Then, under Assumptions* 3.2.2, 3.2.4, 3.3.2, *and* 3.4.1, *the Hamiltonian is minimized with respect to the control $u(t)$ at $u^o(t)$, viz.,*

$$H(x^o(t), u^o(t), \lambda(t), t) \leq H(x^o(t), u(t), \lambda(t), t) \quad \forall\ t \text{ in } [t_0, t_f], \tag{3.103}$$

where

$$-\dot{\lambda}^T(t) = H_x(x^o(t), u^o(t), \lambda(t), t), \quad \lambda^T(t_f) = \phi_x(x^o(t_f)). \tag{3.104}$$

Remark 3.4.1 *Equation* (3.103) *is classically known as the Weierstrass condition.*

Remark 3.4.2 *If a minimum of $H(x^o(t), u(t), \lambda(t), t)$ is found through* (3.103) *at $u^o(t)$ and $H(x^o(t), u(t), \lambda(t), t)$ is twice continuously differentiable with respect to u, then necessary conditions for optimality are*

$$H_u(x^o(t), u^o(t), \lambda(t), t) = 0 \quad \textit{and} \quad H_{uu}(x^o(t), u^o(t), \lambda(t), t) \geq 0. \tag{3.105}$$

The classical Legendre–Clebsch condition is $H_{uu} \geq 0$ and $H_{uu} > 0$ is the strong form.

3.4.1 Control Constraints: Strong Pontryagin Minimum Principle

So far, we have allowed the control $u(t)$ to take on any real value. However, in many (engineering) applications the size of the controls that can be applied is limited. Specifically, we may have bounded control functions in the class

$$u(t) \in \mathcal{U}_B \quad \forall\, t \text{ in } [t_0, t_f], \tag{3.106}$$

where $\mathcal{U}_B$ is a subset of m-dimensional bounded control functions. For example, we may have the scalar control $u(t)$ bounded according to

$$-1 \le u(t) \le 1 \quad \forall\, t \text{ in } [t_0, t_f], \tag{3.107}$$

where the set $\mathcal{U}_B$ is defined as

$$\mathcal{U}_B = \{u(\cdot) \,:\, -1 \le u(\cdot) \le 1\}\,. \tag{3.108}$$

Referring to (3.101), for example, we see that the only change required in our derivation when $u(t) \in \mathcal{U}_B$ is that $u^o(t) \in \mathcal{U}_B$ and that $\eta(\cdot)$ is chosen so that $u^o(t) + \eta(t) \in \mathcal{U}_B$ for all t in $[t_0, t_f]$. Therefore, $u^o(t)$ is found for $u(t) \in \mathcal{U}_B$ which

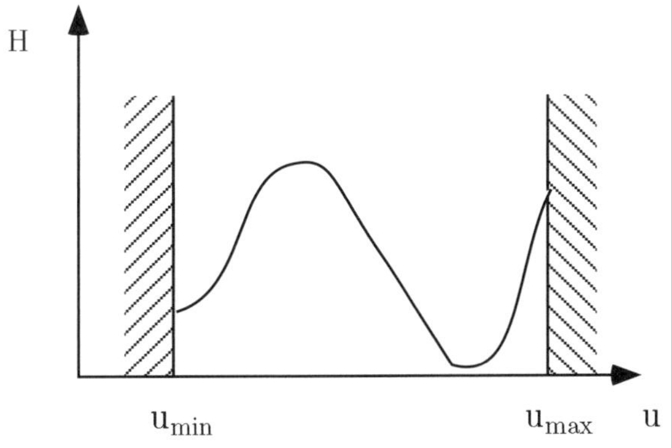

Figure 3.2: Bounded control.

minimizes H. An example of H versus u for bounded control is shown in Figure 3.2. With this modification we have proved the following necessary condition for optimality of $u^o(\cdot)$.

Theorem 3.4.2 *Suppose that $u^o(t) \in \mathcal{U}_B$, $t \in [t_0, t_f]$, minimizes the performance criterion* (3.2) *subject to the dynamic system* (3.41) *and the constraint that $u(t) \in \mathcal{U}_B$ for all $t \in [t_0, t_f]$, and that $H(x(t), u(t), \lambda(t), t)$ is defined according to* (3.52). *Then, under Assumptions* 3.2.2, 3.2.4, 3.3.2, *and* 3.4.1, *the Hamiltonian is minimized with respect to the control $u(t)$, subject to the control constraint $u(t) \in \mathcal{U}_B$, at $u^o(t)$, viz.,*

$$H(x^o(t), u^o(t), \lambda(t), t) \leq H(x^o(t), u(t), \lambda(t), t) \ \forall\ t \ \text{in} \ [t_0, t_f], \tag{3.109}$$

where

$$u^o(t), u(t) \in \mathcal{U}_B \tag{3.110}$$

and

$$-\dot{\lambda}^T(t) = H_x(x^o(t), u^o(t), \lambda(t), t), \quad \lambda^T(t_f) = \phi_x(x^o(t_f)). \tag{3.111}$$

Remark 3.4.3 *Remark* 3.3.1 *notes that if $u(\cdot) \in U_c$, time does not appear explicitly in the Hamiltonian, and the condition $H_u = 0$ holds, the Hamiltonian is constant along the optimal trajectory. Now let $u(\cdot) \in U_{cB}$, the class of piecewise continuously differentiable and bounded functions. The optimal control may be not differentiable at a finite number of points because either it goes from being unconstrained to being on its bound or it is discontinuous by jumping between two (or more) equal minima of H. The fact that $u^o(\cdot) \in U_{cB}$ guarantees that $x(t)$ and $\lambda(t)$ are continuous for all $t \in [t_0, t_f]$. Hence, even if the optimal control is not differentiable at, e.g., point t_d, $x^o(t_d^-) = x^o(t_d^+)$, $\lambda^o(t_d^-) = \lambda^o(t_d^+)$. Consequently, since $u^0(\cdot)$ is chosen to minimize H at any time, $H(x^o(t_d^-), \lambda^o(t_d^-), u^o(t_d^-)) = H(x^o(t_d^+), \lambda^o(t_d^+), u^o(t_d^+))$, i.e., is continuous* *also across t_d. Therefore, for constant control bounds, if H is not an explicit function of time, it remains constant when the control is differentiable, as discussed in Remark 3.3.1. The continuity of the H across the nondifferentiable points of $u^o(\cdot)$ implies that the Hamiltonian remains constant along the entire optimal solution. In this case, it is often referred to as the constant of motion.*

Strong Pontryagin Minimum Principle: Special Case

Let us suppose that we wish to control the linear dynamic system

$$\dot{x}(t) = Ax(t) + Bu(t), \quad x(t_0) = x_0, \tag{3.112}$$

so as to minimize a linear function of the final value of x, viz.,

$$\min_u \alpha^T x(t_f), \tag{3.113}$$

subject to the constraint

$$-1 \le u_i(t) \le 1, \quad i = 1, \dots, m, \tag{3.114}$$

where t_f is the known final time. Here, A is an $n \times n$ constant matrix, B is an $n \times m$ constant matrix, and α is an n-dimensional column vector. We make the following assumption.

Assumption 3.4.2 *The matrices $[B_i, AB_i, \dots, A^{n-1}B_i]$, $i = 1, \dots, m$, have rank n. This is a controllability assumption, where the system is controllable from each control (see [8]).*

Under this assumption we have the following result.

Theorem 3.4.3 *The controls*

$$u_i(t) = -sign[B_i^T e^{-A^T(t-t_f)}\alpha], \quad i = 1, \dots, m, \tag{3.115}$$

are well defined and minimize the performance criterion $\alpha^T x(t_f)$ *in the class of piecewise continuous controls that satisfy* (3.114). *Here,*

$$sign(\sigma) = \begin{cases} 1 & if \;\; \sigma \geq 0, \\ -1 & if \;\; \sigma < 0. \end{cases} \tag{3.116}$$

Proof: First, we show that Assumption 3.4.2 ensures that $B_i^T e^{-A^T(t-t_f)}\alpha$ is not identically zero on a nonzero interval of time so that the control (3.115) is nonzero except possibly at a finite number of times that the control switches as σ goes through zero (3.116). Suppose the converse, i.e., that this function is zero on an interval of time. Then in this interval, at time t say, its time derivatives must also be zero, yielding

$$\begin{aligned} B_i^T e^{-A^T(t-t_f)}\alpha &= 0, \\ B_i^T A^T e^{-A^T(t-t_f)}\alpha &= 0, \\ &\vdots \\ B_i^T A^{T^{n-1}} e^{-A^T(t-t_f)}\alpha &= 0. \end{aligned} \tag{3.117}$$

As the exponential function is always nonsingular and as $\alpha \neq 0$, these equations imply that the rank of $[B_i, AB_i, \dots, A^{n-1}B_i]$ is less than n—a contradiction.

Now we show that the controls (3.115) satisfy Pontryagin's Principle. We have

$$H(x(t), u(t), \lambda(t), t) = \lambda^T(t)Ax(t) + \lambda^T(t)Bu(t), \tag{3.118}$$

so that

$$-\dot{\lambda}^T(t) = \lambda^T(t)A, \quad \lambda^T(t_f) = \alpha^T. \tag{3.119}$$

This equation can be integrated explicitly to yield

$$\lambda(t) = e^{-A^T(t-t_f)}\alpha, \tag{3.120}$$

so that

$$H(x(t), u(t), \lambda(t), t) = \alpha^T e^{-A(t-t_f)}(Ax(t) + Bu(t)), \tag{3.121}$$

and H is minimized by choosing

$$u_i(t) = -sign\left[B_i^T e^{-A^T(t-t_f)}\alpha\right], \quad i = 1, \ldots, m, \tag{3.122}$$

which is just (3.115). So we have proved that (3.115) satisfies Pontryagin's Principle. However, Pontryagin's Principle is only a necessary condition for optimality, so we have not yet proved the optimality of the control. It turns out that this can be verified by direct calculation (cf. Section 3.2.6), but we leave this exercise to the reader. ∎

Remark 3.4.4 *Using* (3.122) *Assumption* 3.4.2 *shows that* $u_i(t) \neq 0$ *for any finite period.*

3.5 Sufficient Conditions for Global Optimality: The Hamilton–Jacobi–Bellman Equation

Throughout, we have stated and proved the fact that Pontryagin's Principle is a necessary condition for optimality. It turns out that only in special cases (cf. Sections 3.2.6 and 3.4.1) it is sufficient for optimality. This should not surprise the reader since in all our derivations we have chosen control perturbations that cause only a small change in $x(\cdot)$ away from $x^o(\cdot)$. In order to derive a sufficient condition for optimality via variational (perturbational) means, one would have to allow arbitrary (not only small) deviations from the trajectory $x^o(\cdot)$. If arbitrary expansions were used, this would involve higher-order derivatives of L, f, and ϕ and would be very cumbersome. Therefore, in this section we take a different approach to sufficiency via the Hamilton–Jacobi–Bellman (H-J-B) equation.

Our approach to global sufficiency centers on the H-J-B partial differential equation

$$-V_t(x(t),t) = \min_{u(t)\in\mathcal{U}_B} [L(x(t),u(t),t) + V_x(x(t),t)f(x(t),u(t),t)] \tag{3.123}$$

with

$$V(x(t_f),t_f) = \phi(x(t_f)). \tag{3.124}$$

The function $V(x(t),t) : \mathbb{R}^n \times \mathbb{R} \to \mathbb{R}$ is known as the optimal value function, and as we will see, $V(x(t),t)$ is the optimal value of the cost criterion starting at $x(t),t$ and using the optimal control to the terminal time. Note that $x(t)$ and t are independent variables in the first-order partial differential equation (3.123) and $V(x(t),t)$ is the dependent variable. The notation $x(t)$ should be read as x at time t. Therefore, $x(t_f)$ in (3.124) is read as x at time t_f. When we consider integration or differentiation along a path, $x(t)$ is dependent on t. Although this equation is rather formidable at first glance, it turns out to be naturally useful in deducing sufficient conditions for optimality and for solving certain classes of optimal control problems.[5] Note that inequality constraints included on the control as well as on the state space become boundary conditions on this first-order partial differential equation. State space constraints are beyond the scope of this book and are not discussed herein.

To develop some facility with the H-J-B partial differential equation, we apply it to the example treated in Section 3.2.6. Here

$$L(x(t),u(t),t) = \frac{1}{2}u^2(t), \tag{3.125}$$

$$f(x(t),u(t),t) = \begin{bmatrix} x_2(t) \\ u(t) \end{bmatrix}, \tag{3.126}$$

$$\phi(x(t_f)) = x_1(t_f). \tag{3.127}$$

[5]The implementation of the H-J-B equation is sometimes referred to as dynamic programming.

Substituting these expressions into (3.123) we obtain

$$\begin{aligned} -V_t(x(t),t) &= \min_{u(t)\in\mathcal{U}}\left[\frac{1}{2}u^2(t) + V_{x_1}(x(t),t)x_2(t) + V_{x_2}(x(t),t)u(t)\right], \\ V(x(t_f),t_f) &= x_1(t_f). \end{aligned} \tag{3.128}$$

Carrying out the minimization of the quantity in square brackets yields

$$u^o(t) = -V_{x_2}(x(t),t), \tag{3.129}$$

and substituting this back into (3.128) yields

$$-V_t(x(t),t) = V_{x_1}(x(t),t)x_2(t) - \frac{1}{2}\left[V_{x_2}(x(t),t)\right]^2. \tag{3.130}$$

We assume that (3.130) has a solution

$$V(x(t),t) = x_1(t) + (t_f - t)x_2(t) - \frac{1}{6}(t_f - t)^3. \tag{3.131}$$

Then

$$\begin{aligned} V_{x_1}(x(t),t) &= 1, \\ V_{x_2}(x(t),t) &= t_f - t, \end{aligned} \tag{3.132}$$

and

$$-V_t(x(t),t) = x_2(t) - \frac{1}{2}(t_f - t)^2. \tag{3.133}$$

Consequently, (3.132) and (3.133) show that (3.131) satisfies (3.128) and, moreover, evaluated at $t = t_f$ equals $x_1(t_f)$. What is more, we have from (3.129) and (3.132) that

$$u^o(t) = -(t_f - t) \tag{3.134}$$

and from (3.131)

$$V(x(t_0),t_0) = x_{10} + (t_f - t_0)x_{20} - \frac{1}{6}(t_f - t_0)^3, \tag{3.135}$$

which agrees with (3.38).

It is clear from these calculations that the H-J-B equation (3.129) is closely related to the optimal solution of the example in Section (3.2.6). We now develop this point fully via the following theorem.

Theorem 3.5.1 *Suppose there exists a once continuously differentiable scalar function* $V(\cdot,\cdot)$ *of* x *and* t *that satisfies the H-J-B equation* (3.123)*. Suppose further that the control* $u^o(x(t),t)$ *that minimizes*

$$L(x(t),u(t),t) + V_x(x(t),t)f(x(t),u(t),t) \tag{3.136}$$

subject to the constraint that $u(t) \in \mathcal{U}_B$*. Then, under the further Assumptions* 3.2.2 *and* 3.3.2 *and that* $f(\cdot,\cdot,\cdot)$, $L(\cdot,\cdot,\cdot)$, *and* $\phi(\cdot)$ *are continuous in all their arguments, the control function* $u^o(x^o(\cdot),\cdot)$ *minimizes* (3.2) *subject to* (3.41) *and* (3.106), *and* $V(x_0,t_0)$ *is equal to the minimum value of* (3.2).

Proof: Under the stated assumptions we note that the identity

$$V(x(t_0),t_0) - V(x(t_f),t_f) + \int_{t_0}^{t_f} \frac{d}{dt}V(x(t),t)dt = 0 \tag{3.137}$$

holds for all piecewise continuous functions $u(\cdot) \in \mathcal{U}_B$. Adding this identically zero quantity to J and noting that

$$\frac{d}{dt}V(x(t),t) = V_t(x(t),t) + V_x(x(t),t)f(x(t),u(t),t) \tag{3.138}$$

yields

$$\begin{aligned} \hat{J}(u(\cdot);x_0) &= V(x_0,t_0) - V(x(t_f),t_f) + \phi(x(t_f)) \\ &\quad + \int_{t_0}^{t_f} [L(x(t),u(t),t) + V_t(x(t),t) \\ &\quad + V_x(x(t),t)f(x(t),u(t),t)]\, dt. \end{aligned} \tag{3.139}$$

Suppose that $V(x(t), t)$ satisfies (3.123). Then

$$\begin{aligned}\hat{J}(u(\cdot); x_0) &= V(x_0, t_0) + \int_{t_0}^{t_f} [\mathcal{H}(x(t), u(t), V_x(x(t), t), t) \\ &\quad - \mathcal{H}(x(t), u^o(x(t), t), V_x(x(t), t), t)]\, dt, \end{aligned} \tag{3.140}$$

where

$$\mathcal{H}(x(t), u(t), V_x(x(t), t), t) \triangleq L(x(t), u(t), t) + V_x(x(t), t) f(x(t), u(t), t). \tag{3.141}$$

From the minimization in (3.123), the integrand of (3.140) is nonnegative and takes on its minimum value of zero when $u(t) = u^o(x(t), t)$. Assumption 3.3.2 ensures that the trajectory resulting from this control is well defined. Consequently, this completes the proof of the theorem. ∎

Remark 3.5.1 *Some interpretation of Equation* (3.140) *is required. Note that the integral is taken along a nonoptimum path generated by* $u(t)$. *If Equation* (3.140) *is rewritten as*

$$\begin{aligned}\Delta J &\triangleq \hat{J}(u(\cdot); x_0) - V(x_0, t_0) \\ &= \int_{t_0}^{t_f} [\mathcal{H}(x(t), u(t), V_x(x(t), t), t) - \mathcal{H}(x(t), u^o(x(t), t), V_x(x(t), t), t)] dt \\ &\triangleq \int_{t_0}^{t_f} \Delta \mathcal{H} dt, \end{aligned} \tag{3.142}$$

then the integral represents the change in cost away from the optimal path. This integral is called Hilbert's integral in the classical calculus of variations literature.

Example 3.5.1 *Consider Hilbert's integral given in* (3.142). *Furthermore, consider also the linear quadratic problem*

$$L(x(t), u(t), t) = \frac{1}{2}[x(t)^T Q x(t) + u(t)^T R u(t)], \tag{3.143}$$

$$f(x(t), u(t), t) = Ax(t) + Bu(t) \tag{3.144}$$

with $Q = Q^T$, $R = R^T > 0$, $\phi(x(t_f)) = 0$. *Then*

$$\Delta\mathcal{H} = \frac{1}{2}[u^T(t)Ru(t) - u^{oT}(x(t),t)Ru^o(t)] + V_x B(u(t) - u^o(x(t),t)). \tag{3.145}$$

Assume the form of $V(\cdot,\cdot)$ *as*

$$V(x(t),t) = \frac{1}{2}x^T(t)S(t)x(t), \quad V_x(x(t),t) = x^T(t)S(t), \tag{3.146}$$

$$u^o(x(t),t) = -R^{-1}B^T S(t)x(t), \quad S(t) = S^T(t) > 0, \tag{3.147}$$

so

$$\begin{aligned}
\Delta\mathcal{H} &= \frac{1}{2}[u^T(t)Ru(t) - u^{oT}(x(t),t)Ru^o(x(t),t)] \\
&\quad + V_x(x(t),t)BR^{-1}R(u(t) - u^o(x(t),t)) \\
&= \frac{1}{2}u^T(t)Ru(t) - \frac{1}{2}u^{oT}(x(t),t)Ru^o(x(t),t) \\
&\quad + u^{oT}(x(t),t)Ru^o(x(t),t) - u^{oT}(x(t),t)Ru(t) \\
&= \frac{1}{2}(u(t) - u^o(x(t),t))^T R(u(t) - u^o(x(t),t)) \geq 0.
\end{aligned} \tag{3.148}$$

This means that $u^o(x^o(t),t)$ *is a global minimum for this linear problem with quadratic performance index.*

Theorem 3.5.1 shows that satisfaction of the H-J-B equation implies that the minimum value of the cost criterion starting at the point $(x(t),t)$ and using the optimal control to the terminal time t_f is equal to $V(x(t),t)$, the optimal value function. To develop a conceptual notion of $V(x(t),t)$, let us consider an optimization problem involving only a terminal cost function, i.e., $J = \phi(x(t_f))$. Suppose there exists an optimal trajectory emanating from $(x(t),t)$. Then, at every point along this path, the value of the optimal value function equals the optimal value $\phi(x^o(t_f))$. Therefore, $V(x(t),t)$ is constant along the motion, and its derivative along the path, assuming continuous first partial derivatives of $V(x(t),t)$, is

$$\frac{dV(x(t),t)}{dt} = \frac{\partial V(x(t),t)}{\partial x(t)} f(x(t),u^o(x(t),t),t) + \frac{\partial V(x(t),t)}{\partial t} = 0, \tag{3.149}$$

where $u^o(x(t),t)$ is the optimal control. If any other control is used, then

$$\frac{dV(x(t),t)}{dt} \geq 0, \tag{3.150}$$

and therefore (3.149) can be restated as

$$-\frac{\partial V(x(t),t)}{\partial t} = \min_{u \in \mathcal{U}_B} \frac{\partial V(x(t),t)}{\partial x(t)} f(x(t),u(t),t). \tag{3.151}$$

This equation can be generalized to that of Equation (3.123) by making the simple transformation that

$$x_{n+1}(t) = \int_{t_0}^{t} L(x(\tau),u(\tau),\tau)d\tau, \tag{3.152}$$

where $x_{n+1}(\cdot)$ is an element of a new state vector

$$\tilde{x}(\cdot) \triangleq [x(\cdot)^T, x_{n+1}(\cdot)]^T \in \mathbb{R}^{n+1}. \tag{3.153}$$

The new terminal cost criterion is

$$\tilde{\phi}(\tilde{x}(t_f)) = \phi(x(t_f)) + x_{n+1}(t_f), \tag{3.154}$$

where

$$\dot{\tilde{x}}(t) = \tilde{f}(x(t),u(t),t) \triangleq \begin{bmatrix} f(x(t),u(t),t) \\ L(x(t),u(t),t) \end{bmatrix}, \quad \tilde{x}(t_0) = \begin{bmatrix} x_0 \\ 0 \end{bmatrix}. \tag{3.155}$$

The H-J-B equation is now

$$\begin{aligned} -V_t(\tilde{x}(t),t) &= \min_{u \in \mathcal{U}_B} V_{\tilde{x}}(\tilde{x}(t),t)\tilde{f}(x(t),u(t),t) \\ &= \min_{u \in \mathcal{U}_B} \Big[V_x(\tilde{x}(t),t) f(x(t),u(t),t) + V_{x_{n+1}}(\tilde{x}(t),t) L(x(t),u(t),t) \Big]. \end{aligned} \tag{3.156}$$

Since the dynamics are not an explicit function of $x_{n+1}(\cdot)$, a variation of $x_{n+1}(\cdot)$ will not change the optimal solution but is just a simple translation in the value of the optimal cost and hence of the optimal value function $V(\tilde{x}(t), t)$. Therefore, $V_{x_{n+1}}(\tilde{x}(t), t) = 1$, and with the notational change, $V(x(t), x_{n+1}(t), t) = V(x(t), t)$, we arrive at Equation (3.123).

We now illustrate Theorem 3.5.1 by applying it to the special case of Section 3.4.1. Here

$$\begin{aligned} L(x(t), u(t), t) &= 0, \\ f(x(t), u(t), t) &= Ax(t) + Bu(t), \\ \mathcal{U}_B &= \{u : -1 \le u_i \le 1, \quad i = 1, \dots, m\}, \\ \phi(x(t_f)) &= \alpha^T x(t_f). \end{aligned} \tag{3.157}$$

The H-J-B equation is

$$-V_t(x(t), t) = \min_{u(t) \in \mathcal{U}_B} \left[V_x(x(t), t) A x(t) + V_x(x(t), t) B u(t) \right], \tag{3.158}$$

which yields

$$u_i^o(x(t), t) = -sign\left[B_i^T V_x^T(x(t), t) \right], \quad i = 1, \dots, m, \tag{3.159}$$

and

$$\begin{aligned} -V_t(x(t), t) &= V_x(x(t), t) A x(t) - \sum_{i=1}^{m} |V_x(x(t), t) B_i|, \\ V(x(t_f), t_f) &= \alpha^T x(t_f). \end{aligned} \tag{3.160}$$

The choice

$$V(x(t), t) \triangleq \alpha^T e^{-A(t - t_f)} x(t) - \int_t^{t_f} \sum_{i=1}^{m} \left| \alpha^T e^{-A(\tau - t_f)} B_i \right| d\tau \tag{3.161}$$

yields

$$\begin{aligned} V_x(x(t), t) &= \alpha^T e^{-A(t - t_f)}, \\ V_t(x(t), t) &= -\alpha^T e^{-A(t - t_f)} A x(t) + \sum_{i=1}^{m} \left| \alpha^T e^{-A(t - t_f)} B_i \right|, \\ V(x(t_f), t_f) &= \alpha^T x(t_f), \end{aligned} \tag{3.162}$$

where, in the expression for $V_t(x(t), t)$, we use the fact that A and $e^{-A(t-t_f)}$ commute. Substituting the expression for $V_x(x(t), t)$ into the right-hand side of (3.160) yields

$$\alpha^T e^{-A(t-t_f)} A x(t) - \sum_{i=1}^{m} \left|\alpha^T e^{-A(t-t_f)} B_i\right|, \tag{3.163}$$

which by (3.162) is $-V_t(x(t), t)$, thus verifying that the H-J-B equation is satisfied by (3.161).

In view of Theorem 3.5.1 we then conclude that the controls given by (3.115) are minimizing and that the minimum value of the performance criterion $\alpha^T x(t_f)$ is given by (3.161) evaluated at $t = t_0$.

Naturally, in more complicated nonlinear control problems, it is more difficult to generate or guess a V function that satisfies the H-J-B equation. Nevertheless, if such a function can be found, the control problem is then completely (globally) solved. It will not have escaped the reader that there is a striking similarity between $V_x(x(t), t)$ and $\lambda^T(t)$ of Pontryagin's Principle. We will develop this relationship in the next subsection.

3.5.1 Derivatives of the Optimal Value Function

In certain classes of problems it is possible to show that when Theorem 3.5.1 holds, the gradient (derivative) with respect to $x(t)$ of the optimal value function $V(x(t), t)$ satisfies the same equation as that for $\lambda^T(t)$ and thereby gives a geometrical interpretation to the Lagrange multipliers. Here we illustrate this when $\mathcal{U}$ is the whole m-dimensional space, viz., $\mathbb{R}^m$. This derivation assumes that the second partial derivative of $V(x(t), t)$ with respect to $x(t)$ exists. Furthermore, there are no control constraints and (3.123) becomes

$$\begin{aligned} -V_t(x(t), t) &= \min_{u \in \mathcal{U}} \left[L(x(t), u(t), t) + V_x(x(t), t) f(x(t), u(t), t)\right], \\ V(x(t_f), t_f) &= \phi(x(t_f)). \end{aligned} \tag{3.164}$$

Since we take time derivatives along the optimal path generated by the optimal control, we replace the arbitrary initial state $x(t)$ in the H-J-B equation with $x^o(t)$. Calling the minimizing control $u^o(x^o(t),t)$, we obtain

$$\begin{aligned} -V_t(x^o(t),t) &= [L(x^o(t),u^o(x^o(t),t),t) + V_x(x^o(t),t)f(x^o(t),u^o(x^o(t),t),t)] \\ &= \mathcal{H}(x^o(t),u^o(x^o(t),t),V_x(x^o(t),t),t). \end{aligned} \tag{3.165}$$

First Derivative of the Optimal Value Function

Assuming that the necessary derivatives exist, we differentiate (3.165) with respect to $x^o(t)$ to obtain

$$\begin{aligned} -V_{tx}(x^o(t),t) &= L_x(x^o(t),u^o(x^o(t),t),t) \\ &\quad + V_x(x^o(t),t)f_x(x^o(t),u^o(x^o(t),t),t) \\ &\quad + f^T(x^o(t),u^o(x^o(t),t),t)V_{xx}(x^o(t),t) \\ &\quad + L_u(x^o(t),u^o(x^o(t),t),t)u^o_x(x^o(t),t) \\ &\quad + V_x(x^o(t),t)f_u(x^o(t),u^o(x^o(t),t),t)u^o_x(x^o(t),t). \end{aligned} \tag{3.166}$$

Note that

$$V_{tx}(x^o(t),t) = V_{xt}(x^o(t),t) \tag{3.167}$$

if the second partial derivatives of V exist, and

$$\frac{d}{dt}V_x(x^o(t),t) = V_{xt}(x^o(t),t) + f^T(x^o(t),u^o(x^o(t),t),t)V_{xx}(x^o(t),t). \tag{3.168}$$

Using the equations in (3.166) and noting that, because of the minimization with respect to $u(t)$,

$$L_u(x^o(t),u^o(x^o(t),t),t) + V_x(x^o(t),t)f_u(x^o(t),u^o(x^o(t),t),t) = 0, \tag{3.169}$$

we obtain

$$-\dot{V}_x(x^o(t),t) = L_x(x^o(t),u^o(x^o(t),t),t) + V_x(x^o(t),t)f_x(x^o(t),u^o(x^o(t),t),t). \quad (3.170)$$

It is also clear from (3.164) that

$$V_x(x(t_f),t_f) = \phi_x(x(t_f)). \quad (3.171)$$

Comparing (3.170) and (3.171) to (3.56) and (3.57), we see that $V_x(x^o(t),t)$ satisfies the same differential equation and has the same final value as $\lambda(t)^T$, which can be interpreted as the derivative of the optimal value function with respect to the state.

There are, however, control problems where there is no continuously differentiable $V(\cdot,\cdot)$ that satisfies the H-J-B equation, but Pontryagin's $\lambda(\cdot)$ is well defined by its differential equation. In such cases, it is awkward to find a physical interpretation for $\lambda(\cdot)$.

Second Derivative of the Optimal Value Function

The second derivative of the optimal value function is now derived and is, in general, the curvature of the optimal value function with respect to the state. Here, certain assumptions must be made first for simplicity of the derivation.

Assumption 3.5.1 $u^o(x(\cdot),\cdot) \in$ Interior $\mathcal{U}$.

If the optimal control function does not lie on a control boundary, then there are no restrictions assuming an arbitrarily small control variation away from the optimal control. (If u is on a bound, then $u(x(t),t) = u(t) \Rightarrow u_x(x(t),t) = 0$. Using this the following results would have to be modified.)

Assumption 3.5.2

$$\mathcal{H}(x^o(t),u^o(x^o(t),t),V_x(x^o(t),t),t) < \mathcal{H}(x^o(t),u(t),V_x(x^o(t),t),t) \quad (3.172)$$

implies

$$\begin{aligned}
\mathcal{H}_u(x^o(t), u^o(x^o(t), t), V_x(x^o(t), t), t) &= 0, \\
\mathcal{H}_{uu}(x^o(t), u^o(x^o(t), t), V_x(x^o(t), t), t) &> 0.
\end{aligned} \tag{3.173}$$

Then from the H-J-B equation (3.165)

$$\begin{aligned}
-V_{txx}(x^o(t), t) &= \mathcal{H}_{xx}(x^o(t), u^o(x^o(t), t), V_x(x^o(t), t), t) \\
&\quad + \mathcal{H}_{xu}(x^o(t), u^o(x^o(t), t), V_x(x^o(t), t), t) u^o_x(x^o(t), t) \\
&\quad + \mathcal{H}_{xV_x}(x^o(t), u^o(x^o(t), t), V_x(x^o(t), t), t) V_{xx}(x^o(t), t) \\
&\quad + V_{xx}(x^o(t), t) \mathcal{H}_{V_x x}(x^o(t), u^o(x^o(t), t), V_x(x^o(t), t), t) \\
&\quad + V_{xx}(x^o(t), t) \mathcal{H}_{V_x u}(x^o(t), u^o(x^o(t), t), V_x(x^o(t), t), t) u^o_x(x^o(t), t) \\
&\quad + \sum_i^n \mathcal{H}_{V_{x_i}}(x^o(t), u^o(x^o(t), t), V_x(x^o(t), t), t) V_{x_i xx}(x^o(t), t) \\
&\quad + u^{oT}_x(x^o(t), t) \mathcal{H}_{ux}(x^o(t), u^o(x^o(t), t), V_x(x^o(t), t), t) \\
&\quad + u^{oT}_x(x^o(t), t) \mathcal{H}_{uV_x}(x^o(t), u^o(x^o(t), t), V_x(x^o(t), t), t) V_{xx} \\
&\quad + u^{oT}_x(x^o(t), t) \mathcal{H}_{uu}(x^o(t), u^o(x^o(t), t), V_x(x^o(t), t), t) u^o_x(x^o(t), t),
\end{aligned} \tag{3.174}$$

where derivatives of $\mathcal{H}(x^o(t), u^o(x^o(t), t), V_x(x^o(t), t), t)$ are taken with respect to each of its explicit arguments. It is assumed here that third derivatives exist. Furthermore, the function $\mathcal{H}$ is used to avoid tensor products, i.e., taking second partials of a dynamic vector function $f(x(t), u^o(x(t), t), t)$ with respect to the state vector. The boundary condition for (3.174) is

$$V_{xx}(x(t_f), t_f) = \phi_{xx}(x(t_f)). \tag{3.175}$$

Since a differential equation for V_{xx} along the optimal path is being sought, V_{xx} is directly differentiated as

$$\frac{dV_{xx}(x^o(t),t)}{dt}$$
$$= \sum_{i}^{n} \mathcal{H}_{V_{x_i}}(x^o(t), u^o(x^o(t),t), V_x(x^o(t),t),t)V_{x_i xx}(x^o(t),t) + V_{xxt}(x^o(t),t)$$
$$= \sum_{i}^{n} f_i(x^o(t), u^o(x^o(t),t), V_x(x^o(t),t),t)V_{x_i xx}(x^o(t),t) + V_{xxt}(x^o(t),t), \quad (3.176)$$

where the sum is used to construct the tensor product. Substitution of Equation (3.174) into (3.176) gives the differential equation

$$\begin{aligned}
-\frac{dV_{xx}(x^o(t),t)}{dt} &= \mathcal{H}_{xx}(x^o(t), u^o(x^o(t),t), V_x(x^o(t),t),t) \\
&+ \mathcal{H}_{xu}(x^o(t), u^o(x^o(t),t), V_x(x^o(t),t),t)u^o_x(x^o(t),t) \\
&+ \mathcal{H}_{xV_x}(x^o(t), u^o(x^o(t),t), V_x(x^o(t),t),t)V_{xx}(x^o(t),t) \\
&+ V_{xx}(x^o(t),t)\mathcal{H}_{V_x x}(x^o(t), u^o(x^o(t),t), V_x(x^o(t),t),t) \\
&+ V_{xx}(x^o(t),t)\mathcal{H}_{V_x u}(x^o(t), u^o(x^o(t),t), V_x(x^o(t),t),t)u^o_x(x^o(t),t) \\
&+ u^{oT}_x(x^o(t),t)\mathcal{H}_{ux}(x^o(t), u^o(x^o(t),t), V_x(x^o(t),t),t) \\
&+ u^{oT}_x \mathcal{H}_{uV_x}(x^o(t), u^o(x^o(t),t), V_x(x^o(t),t),t)V_{xx} \\
&+ u^{oT}_x(x^o(t),t)\mathcal{H}_{uu}(x^o(t), u^o(x^o(t),t), V_x(x^o(t),t),t)u^o_x(x^o(t),t).
\end{aligned} \quad (3.177)$$

Finally, an expression for $u^o_x(x^o(t),t)$ is required. From Assumptions 3.5.1 and 3.5.2, it follows that

$$\mathcal{H}_u(x^o(t), V_x(x^o(t),t), u^o(x^o(t),t),t) = 0 \quad (3.178)$$

is true for all $x^o(t)$, and then

$$\begin{aligned}
&\mathcal{H}_{ux}(x^o(t), u^o(x^o(t),t), V_x(x^o(t),t),t) \\
&\quad + \mathcal{H}_{uV_x}(x^o(t), u^o(x^o(t),t), V_x(x^o(t),t),t)V_{xx}(x^o(t),t) \\
&\quad + \mathcal{H}_{uu}(x^o(t), u^o(x^o(t),t), V_x(x^o(t),t),t)u^o_x(x^o(t),t) = 0.
\end{aligned} \quad (3.179)$$

This, given Assumption 3.5.2, produces

$$
\begin{aligned}
u_x^o(x^o(t),t) &= -\mathcal{H}_{uu}^{-1}(x^o(t),u^o(x^o(t),t),V_x(x^o(t),t),t) \\
&\quad \times (\mathcal{H}_{ux}(x^o(t),u^o(x^o(t),t),V_x(x^o(t),t),t) \\
&\quad + \mathcal{H}_{uV_x}(x^o(t),u^o(x^o(t),t),V_x(x^o(t),t),t)V_{xx}(x^o(t),t)). \qquad (3.180)
\end{aligned}
$$

Substitution of Equation (3.180) into (3.177) gives, after some manipulations and removing the arguments,

$$
\begin{aligned}
-\frac{dV_{xx}}{dt} &= (f_x^T - \mathcal{H}_{xu}\mathcal{H}_{uu}^{-1}f_u^T)V_{xx} + V_{xx}(f_x - f_u\mathcal{H}_{uu}^{-1}\mathcal{H}_{ux}) \qquad (3.181)\\
&\quad + (\mathcal{H}_{xx} - \mathcal{H}_{xu}\mathcal{H}_{uu}^{-1}\mathcal{H}_{ux}) - V_{xx}f_u\mathcal{H}_{uu}^{-1}f_u^T V_{xx},
\end{aligned}
$$

where $\mathcal{H}_{V_x x} = f_x$ and $\mathcal{H}_{V_x u} = f_u$. Note that expanding $\mathcal{H}_{xx}$, $\mathcal{H}_{xu}$, and $\mathcal{H}_{uu}$ produces tensor products. The curvature of $V(x^o(t),t)$, $V_{xx}(x^o(t),t)$ along an optimal path is propagated by a Riccati differential equation, and the existence of its solution will play an important role in the development of local sufficiency for a weak minimum. Note that the problem of minimizing a quadratic cost criterion subject to a linear dynamic constraint as given in Example 3.5.1 produces a Riccati differential equation in the symmetric matrix $S(t)$. More results will be given in Chapter 5.

3.5.2 Derivation of the H-J-B Equation

In Theorem 3.5.1 we showed that if there is a solution to the H-J-B equation which satisfies certain conditions, then this solution evaluated at $t = t_0$ is the optimal value of (3.2). Here, we prove the converse, viz., that under certain assumptions on the optimal value function, the H-J-B equation can be deduced.

Let us define

$$
V(x(t),t) \triangleq \min_{u(\cdot)\in\mathcal{U}_B}\left[\phi(x(t_f)) + \int_t^{t_f} L(x(\tau),u(\tau),\tau)d\tau\right], \qquad (3.182)
$$

where $u(\tau) \in \mathcal{U}_B$ for all τ in $[t, t_f]$, and let $u^o(\tau; x(t), t)$, $t \le \tau \le t_f$ be the optimal control function for the dynamic system (3.41) with "initial condition" $x(t)$ at $\tau = t$. Then

$$V(x(t), t) = \phi(x(t_f)) + \int_t^{t_f} L(x(\tau), u^o(\tau; x(t), t), \tau) d\tau, \tag{3.183}$$

so that

$$\frac{d}{dt} V(x(t), t) = -L(x(t), u^o(t; x(t), t), t) \tag{3.184}$$

and

$$V(x(t_f), t_f) = \phi(x(t_f)). \tag{3.185}$$

We can now proceed to the theorem.

Theorem 3.5.2 *Suppose that the optimal value function $V(\cdot, \cdot)$ defined by* (3.184) *and* (3.185) *is once continuously differentiable in $x(t)$ and t. Then $V(\cdot, \cdot)$ satisfies the H-J-B equation* (3.123) *and* (3.124).

Proof: From the existence of $V(\cdot, \cdot)$, (3.184), and the differentiability of $V(\cdot, \cdot)$ we have that

$$-V_t(x(t), t) = V_x(x(t), t) f(x(t), u^o(t; x(t), t), t) + L(x(t), u^o(t; x(t), t), t). \tag{3.186}$$

Furthermore, it follows from (3.182) that

$$\begin{aligned} V(x(t), t) \le & \int_t^{t+\Delta} L(x(\tau), u(\tau), \tau) d\tau \\ & + \int_{t+\Delta}^{t_f} L(x(\tau), u^o(\tau; x(t+\Delta), (t+\Delta)), \tau) d\tau \\ & + \phi(x(t_f)), \end{aligned} \tag{3.187}$$

where $u(\tau) \in \mathcal{U}_B$, $t \leq \tau \leq t+\Delta$, is an arbitrary continuous function for some positive $0 < \Delta \ll t_f - t$. This yields

$$V(x(t),t) \leq \int_t^{t+\Delta} L(x(\tau),u(\tau),\tau)d\tau + V(x(t+\Delta),(t+\Delta)), \qquad (3.188)$$

and expanding the right-hand side in Δ yields

$$\begin{aligned} V(x(t),t) \leq\ & L(x(t),u(t),t)\Delta + V(x(t),t) + V_t(x(t),t)\Delta \\ & + V_x(x(t),t)f(x(t),u(t),t)\Delta + \mathcal{O}(\Delta), \end{aligned} \qquad (3.189)$$

which in turn yields

$$0 \leq [L(x(t),u(t),t) + V_t(x(t),t) + V_x(x(t),t)f(x(t),u(t),t)]\,\Delta + \mathcal{O}(\Delta). \quad (3.190)$$

This inequality holds for all continuous $u(\cdot)$ on $[t, t+\Delta]$ and all $\Delta \geq 0$. Hence, for all $u(t) \in \mathcal{U}_B$

$$0 \leq L(x(t),u(t),t) + V_t(x(t),t) + V_x(x(t),t)f(x(t),u(t),t). \qquad (3.191)$$

Considering (3.186) and (3.191) together we conclude that

$$\min_{u(t)\in\mathcal{U}_B} [V_t(x(t),t) + L(x(t),u(t),t) + V_x(x(t),t)f(x(t),u(t),t)] = 0. \qquad (3.192)$$

Since $V_t(x(t),t)$ does not depend upon $u(t)$, we can rewrite (3.192) as

$$-V_t(x(t),t) = \min_{u(t)\in\mathcal{U}_B} [L(x(t),u(t),t) + V_x(x(t),t)f(x(t),u(t),t)]\,. \qquad (3.193)$$

This, together with (3.185), is just the H-J-B equation (3.123) and (3.124). ▮

Note that the above theorem and proof are based squarely on the assumption that $V(\cdot,\cdot)$ is once continuously differentiable in both arguments. It turns out that there are problems where this smoothness is not present so that this assumption is violated.

In any event, however, it is a nasty assumption as one has to solve the optimal control problem to obtain $V(\cdot,\cdot)$ before one can verify it. Consequently, the above theorem is largely only of theoretical value (it increases one's insight into the H-J-B equation). The main strength of the H-J-B equation lies in Theorem 3.5.1, which provides sufficient conditions for optimality.

In many derivations of the H-J-B equation it is assumed that the second partial derivatives of $V(x(t),t)$ exist. This is done because it is known that in certain classes of optimal control problems optimality is lost (i.e., J can be made arbitrarily large and negative) if $V_{xx}(x(t),t)$ ceases to exist at a certain time t'. However, this is a red herring in the derivation of the H-J-B equation; the expansion in (3.189) is clearly valid if one merely assumes that $V(x(t),t)$ exists and is once continuously differentiable.

3.6 Unspecified Final Time t_f

So far in this chapter we have assumed that the final time t_f is given. One could, however, treat t_f as a parameter and, along with $u(\cdot)$, choose it to minimize (3.2). Clearly, a necessary condition for the optimality of $t_f = t_f^o$ is that

$$J_{t_f}(u^o(\cdot); x_0, t_f^o) = 0, \tag{3.194}$$

where

$$J(u(\cdot); x_0, t_f) \triangleq \phi(x(t_f), t_f) + \int_{t_0}^{t_f} L(x(t), u(t), t)dt, \tag{3.195}$$

provided, of course, that J is differentiable at t_f^o. Equation (3.194) is known as a transversality condition.

Note that in (3.195) we allow a more general $\phi(\cdot,\cdot)$ than previously, viz., one that depends on both $x(t_f)$ and t_f. Naturally, in the case where t_f is given, nothing is gained by this explicit dependence of ϕ on t_f.

Suppose that $u^o(\cdot)$ minimizes (3.195) when $t_f = t_f^o$. Then

$$J(u^o(\cdot); x_0, t_f^o) = \phi(x^o(t_f^o), t_f^o) + \int_{t_0}^{t_f^o} L(x^o(t), u^o(t), t)\,dt \tag{3.196}$$

and

$$J(u^o(\cdot); x_0, t_f^o + \Delta) = \phi(x^o(t_f^o + \Delta), t_f^o + \Delta) + \int_{t_0}^{t_f^o+\Delta} L(x^o(t), u^o(t), t)\,dt, \tag{3.197}$$

where, if $\Delta > 0$, $u^o(\cdot)$ over the interval $(t_f^o, t_f^o + \Delta)$ is any continuous function emanating from $u^o(t_f)$ with values in $\mathcal{U}$.

Subtracting (3.196) from (3.197) and expanding the right-hand side of (3.197) yields

$$\begin{aligned} &J(u^o(\cdot); x_0, t_f^o + \Delta) - J(u^o(\cdot); x_0, t_f^o) \\ &\quad = L(x^o(t_f^o), u^o(t_f^o), t_f^o)\Delta + \phi_{t_f}(x^o(t_f^o), t_f^o)\Delta \\ &\qquad + \phi_x(x^o(t_f^o), t_f^o) f(x^o(t_f^o), u^o(t_f^o), t_f^o)\Delta + \mathcal{O}(\Delta). \end{aligned} \tag{3.198}$$

It follows that for t_f^o to be optimal,

$$\begin{aligned} J_{t_f}(u^o(\cdot); x_0, t_f^o) &= L(x^o(t_f^o), u^o(t_f^o), t_f^o) \\ &\quad + \phi_x(x^o(t_f^o), t_f^o) f(x^o(t_f^o), u^o(t_f^o), t_f^o) + \phi_{t_f}(x^o(t_f^o), t_f^o) = 0. \end{aligned} \tag{3.199}$$

This condition (the so-called transversality condition) can be written using the Hamiltonian H as

$$\phi_{t_f}(x^o(t_f^o), t_f^o) + H(x^o(t_f^o), u^o(t_f^o), \lambda(t_f^o), t_f^o) = 0. \tag{3.200}$$

Since, by (3.109), H is minimized by $u^o(t_f^o)$ and since $x^o(\cdot)$, $\lambda(\cdot)$, $L(\cdot,\cdot,\cdot)$, and $f(\cdot,\cdot,\cdot)$ are continuous functions of time, it is clear that $u^o(t_f^o)$ and $u^o(t_f^{o-})$ both yield the same value of H. It should be noted that jumps are allowed in the control, where identical values of the minimum Hamiltonian H will occur at different values of u at t_f. Consequently, (3.200) holds with $u^o(t_f^o)$ replaced by $u^o(t_f^{o-})$ and because of this it is not necessary to assume for $\Delta < 0$ that $u^o(\cdot)$ is continuous from the left at t_f^o. We have thus proved the following theorem.

Theorem 3.6.1 *Suppose that $\phi(\cdot,\cdot)$ depends explicitly on t_f as in (3.195) and that t_f is unspecified. Then, if the pair $u^o(\cdot)$, t_f^o minimizes (3.195), the condition*

$$\phi_{t_f}(x^o(t_f^o), t_f^o) + H(x^o(t_f^o), u^o(t_f^o), \lambda(t_f^o), t_f^o) = 0 \tag{3.201}$$

holds in addition to (3.109).

Remark 3.6.1 *It should be noted that free final time problems can always be solved by augmenting the state vector with the time by defining a new independent variable and solving the resulting problem as a fixed final "time" problem. This is done by using the transformation*

$$t = (t_f - t_0)\tau + t_0, \tag{3.202}$$

$$\Rightarrow dt = (t_f - t_0)d\tau, \tag{3.203}$$

where τ goes from 0 to 1. The differential constraints become

$$\begin{bmatrix} x' \\ t' \end{bmatrix} = \begin{bmatrix} (t_f - t_0)f(\tau, x, u, t_f) \\ (t_f - t_0) \end{bmatrix}, \tag{3.204}$$

where $x' = dx/d\tau$. We will not use this approach here.

Example 3.6.1 *The simplest example to illustrate (3.201) is the scalar system ($n = m = 1$)*

$$\dot{x}(t) = bu(t), \quad b \neq 0, \quad x(t_0) = x_0, \tag{3.205}$$

with a performance criterion

$$J(u(\cdot); x_0, t_f) = \alpha x(t_f) + \frac{1}{2}t_f^2 + \int_{t_0}^{t_f} \frac{1}{2}u^2(t)\,dt. \tag{3.206}$$

Here

$$H(x(t), u(t), \lambda(t), t) = \frac{1}{2}u(t)^2 + \lambda(t)bu(t) \tag{3.207}$$

so that

$$\lambda(t) = \alpha \tag{3.208}$$

and

$$u^o(t) = -b\alpha. \tag{3.209}$$

From (3.201), *the optimal final time is determined as*

$$t_f^o - \frac{1}{2}b^2\alpha^2 = 0 \quad \Rightarrow \quad t_f^o = \frac{1}{2}b^2\alpha^2. \tag{3.210}$$

The optimal cost criterion is

$$J(u^o(\cdot); x_0, t_f) = \alpha x_0 - \frac{1}{2}\alpha^2 b^2(t_f - t_0) + \frac{1}{2}t_f^2 \tag{3.211}$$

evaluated at t_f^o *given in* (3.210). *Its derivative with respect to* t_f *is zero at* t_f^o, *thus verifying* (3.199).

Problems

1. Extremize the performance index

$$J = \frac{1}{2}\int_0^1 e^{(u-x)^2}\,dt$$

subject to

$$\dot{x} = u, \qquad x(0) = x_0 = 1.$$

(a) Find the state, control, and multipliers as a function of time.

(b) Show that the Hamiltonian is a constant along the motion.

2. Minimize the performance index

$$J = \int_0^2 (|u| - x)\,dt$$

subject to

$$\dot{x} = u, \quad x(0) = 1, \quad |u| \le 1.$$

(a) Determine the optimal state and control histories.

(b) Discuss the derivation of the first-order necessary conditions and their underlying assumption. What modifications in the theory are required for this problem? Are there any difficulties making the extension? Explain.

3. Consider the following control problem with

$$\begin{aligned} \dot{x} &= A(t)x(t) + B(t)u(t), \quad x(t_0) = x_0, \\ \min_{u(t)} J &= \int_{t_0}^{t_f} (a^T x + \frac{1}{2} u^T R u)\,dt, \quad R(t) > 0. \end{aligned}$$

Prove that for this problem, a satisfaction of Pontryagin's Minimum Principle by a control function u^* is a *necessary and sufficient* condition for u^* to be the control that minimizes J.

4. Consider the problem of minimizing with respect to the control $u(\cdot)$ the cost criterion

$$J = \int_{t_0}^{t_f} \frac{1}{2} u^T R u dt + \phi(x(t_f)), \quad R > 0,$$

subject to

$$\dot{x} = Ax + Bu, \quad x(t_0) = x_0 \quad \text{(given)},$$

where $\phi(x(t_f))$ is twice continuously differentiable function of x. Suppose that we require $u(t)$ to be piecewise *constant* as follows:

$$u(t) = \begin{cases} u_0, & t_0 \le t \le t_1, \\ u_1, & t_1 \le t \le t_f, \end{cases}$$

where t_1 is specified and $t_0 < t_1 < t_f$. Derive from first principles the first-order necessary conditions that must be satisfied by an optimal control function

$$u^*(t) = \begin{cases} u_0^*, & t_0 \le t \le t_1, \\ u_1^*, & t_1 \le t \le t_f. \end{cases}$$

State explicitly all assumptions made.

5. Consider the problem of minimizing with respect to $u(\cdot)$ the cost criterion

$$J = \lim_{t_f \to \infty} \int_0^{t_f} \left[\frac{qx^4}{4} + \frac{u^2}{2} \right] dt$$

subject to

$$\dot{x} = u.$$

Find the optimal feedback law. *Hint*: If you use the H-J-B equation to solve this problem, assume that the optimal value function is only an explicit function of x. Check to ensure that all conditions are satisfied for H-J-B theory.

6. Minimize the functional

$$\min_{u(\cdot) \in \mathcal{U}} = \int_0^{t_f} (x^2 + u^2)\, dt$$

subject to

$$\dot{x} = u, \quad x(0) = x_0.$$

(a) Write down the first-order necessary conditions.

(b) Solve the two-point boundary-value problem so that u is obtained as a function of t and x.

(c) Show that the Hamiltonian is a constant along the extremal path.

7. Consider the problem of the Euler–Lagrange equation

$$\min_{u(\cdot)\in\mathcal{U}} \int_0^{t_f} L(x,u)dt,$$

where $u(t) \in \mathbb{R}^m$ and $x(t) \in \mathbb{R}^n$ subject to

$$\dot{x} = u, \quad x(0) = x_0.$$

(a) Show that the first-order necessary conditions reduce to

$$\frac{d}{dt}L_u - L_x = \frac{d}{dt}L_{\dot{x}} - L_x = 0.$$

(b) Show that the Hamiltonian is a constant along the motion.

8. Let

$$T = \frac{1}{2}m\dot{x}^2, \quad U = \frac{1}{2}kx^2,$$

and $L = T - U$.

Use the Euler–Lagrange equations given in problem 7 to produce the equation of motion.

9. Extremize the performance index

$$J = \frac{1}{2}\left[cx_f^2 + \int_{t_0}^{t_f} u^2 dt\right],$$

where c is a positive constant. The dynamic system is defined by

$$\dot{x} = u,$$

and the prescribed boundary conditions are

$$t_0, x_0, t_f \equiv \text{ given.}$$

10. Extremize the performance index

$$J = \frac{1}{2}\int_{t_0}^{t_f} (u^2 - x)\, dt$$

subject to the differential constraint

$$\dot{x} = 0$$

and the prescribed boundary conditions

$$t_0 = 0, \quad x_0 = 0, \quad t_f = 1.$$

11. Extremize the performance index

$$J = \frac{1}{2}\int_{t_0}^{t_f} \left[\frac{(u^2+1)}{x}\right]^{1/2} dt$$

subject to the differential constraint

$$\dot{x} = u$$

and the prescribed boundary conditions

$$t_0 = 0, \quad x_0 > 0, \quad t_f = \text{ given } > 0.$$

12. Minimize the performance index

$$J = bx(t_f) + \frac{1}{2}\int_{t_0}^{t_f} \left\{-[u(t) - a(t)]^2 + u^4\right\} dt$$

subject to the differential constraint

$$\dot{x} = u$$

and the prescribed boundary conditions

$$t_0 = 0, \quad x_0 = 1, \quad a(t) = t, \quad t_f = 1.$$

Find a value for b so that the control jumps somewhere in the interval $[0, 1]$.

13. Minimize analytically and numerically the performance index

$$J = \alpha_1 x_1(t_f)^2 + \alpha_2 x_2(t_f)^2 + \int_{t_0}^{t_f} |u| \, dt$$

subject to

$$\ddot{x} = u, \quad |u| \leq 1,$$

and the prescribed boundary conditions

$$t_0 = 0, \quad x_1(0) = 10, \quad x_2(0) = 0, \quad t_f = 10.$$

Since $L = |u|$ is not continuously differentiable, discuss the theoretical consequences.

14. Minimize the performance index

$$J = \int_0^6 (|u| - x_1) \, dt - 2x_1(6)$$

subject to

$$\begin{aligned} \dot{x}_1 &= x_2, \quad x_1(0) = 0, \\ \dot{x}_2 &= u, \quad x_2(0) = 0, \\ |u| &\leq 1. \end{aligned}$$

Determine the optimal state and control histories.

15. Consider the problem of minimizing with respect to the control $u(\cdot)$ the cost criterion

$$J = \phi(x(t_f)) + \int_{t_0}^{t_f} L(x, u, t) \, dt$$

subject to

$$\dot{x} = f(x, u, t), \quad x(t_0) = x_0 \text{ given,}$$

where F is continuously differentiable in x, and L and f are continuously differentiable in x, u, and t. Suppose that we require $u(t)$ to be piecewise *constant* as follows:

$$u(t) = \begin{cases} u_0, & t_0 \le t \le t_1, \\ u_1, & t_1 \le t \le t_f, \end{cases}$$

where t_1 is specified and $t_0 < t_1 < t_f$. Derive from first principles the first-order necessary conditions that must be satisfied by an optimal control function, i.e.,

$$u^o(t) = \begin{cases} u_0^o, & t_0 \le t \le t_1, \\ u_1^o, & t_1 \le t \le t_f. \end{cases}$$

16. Consider the problem of minimizing

$$\phi(x(t_f)) = \alpha^T x(t_f), \qquad t_f \text{ fixed,}$$

subject to

$$\dot{x} = Ax + Bu$$

for $u \in \mathcal{U} \triangleq \{u : -1 \le u_i \le 1, \ \ i = 1, \dots, m\}$.

(a) Determine the optimal value function $V(x, t)$.

(b) Write the change in cost from optimal using Hilbert's integral and explain how this change in cost can be computed.

17. Consider the problem of minimizing with respect to $u(\cdot) \in \mathcal{U}$

$$J = F(x(t_0) + \phi(x(t_f))$$

subject to

$$\begin{aligned} \dot{x} &= Ax + Bu, \\ \mathcal{U} &= \{u(\cdot) : u(\cdot) \text{ are piecewise continuous functions}\}, \end{aligned}$$

where t_0 is the initial time, and $t_0 < t_f$, with both fixed.

Develop and give the conditions for weak and strong local optimality. List the necessary assumptions.

18. Minimize the performance index

$$J = \int_0^4 (|u| - x)\, dt + 2x(4)$$

subject to

$$\begin{aligned} \dot{x} &= u \\ x(0) &= 0, \\ |u| &\leq 1. \end{aligned}$$

Determine the optimal state and control histories.

19. Derive the H-J-B equation for the optimization problem of minimizing, with respect to $u(\cdot) \in \mathcal{U}$, the cost criterion

$$J = e^{\left[\int_{t_0}^{t_f} L dt + \phi(x(t_f))\right]}$$

subject to $\dot{x} = f(x, u, t)$.

20. Find the control $u(\cdot) \in \mathcal{U}_B \triangleq \{u : |u| \leq 1\}$ that minimizes

$$J = (x_1(50) - 10)^2 q_1 + x_2^2(50) q_2 + \int_0^{50} |u| dt$$

subject to

$$\begin{bmatrix} \dot{x}_1 \\ \dot{x}_2 \end{bmatrix} = \begin{bmatrix} 0 & 1 \\ 0 & 0 \end{bmatrix} \begin{bmatrix} x_1 \\ x_2 \end{bmatrix} + \begin{bmatrix} 0 \\ 1 \end{bmatrix} u, \quad \begin{bmatrix} x_1(0) \\ x_2(0) \end{bmatrix} = \begin{bmatrix} 0 \\ 0 \end{bmatrix}.$$

Choose q_1 and q_2 large enough to approximately satisfy the terminal constraints $x_1(50) = 10$ and $x_2(50) = 0$.

CHAPTER 4

Terminal Equality Constraints

4.1 Introduction

Theorem 3.3.1, a weak form of the Pontryagin Principle, is a condition that is satisfied if $x^o(\cdot)$, $u^o(\cdot)$ are an optimal pair (i.e., a necessary condition for optimality). It is also clear from (3.55) that satisfaction of this condition is sufficient for the change in the performance criterion to be nonnegative, to first order, for any weak perturbation in the control away from $u^o(\cdot)$.

In this chapter we first derive a *weak* form of the Pontryagin Principle as a necessary condition for optimality when *linear* terminal equality constraints are present and when the dynamic system is linear. When nonlinear terminal equality constraints are present and when the dynamic system is nonlinear, the elementary constructions used in this book turn out to be inadequate to deduce necessary conditions; indeed, deeper mathematics is required here for rigorous derivations. If deriving necessary conditions in these more involved control problem formulations is out of our reach, we can, nevertheless, rather easily and rigorously show that Pontryagin's Principle is a *sufficient* condition for "weak or strong first-order optimality." This confirms that if Pontryagin's Principle is satisfied, the change in J can be negative only as a consequence of second- and/or higher-order terms in its expansion. Thus Pontryagin's Principle can be thought of as a first-order optimality condition.

By way of introducing optimal control problems with terminal equality constraints we begin in Section 4.2.1 with the problem of steering a linear dynamic system from an initial point to the origin in specified time while minimizing the control "energy" consumed. It turns out that the most elementary, direct methods are adequate to handle this problem. We then derive a Pontryagin-type necessary condition for optimality for the case of a general performance criterion and both linear and nonlinear terminal equality constraints. Turning to nonlinear terminal equality constraints and nonlinear dynamic systems we introduce the notion of weak first-order optimality and state a weak form of the Pontryagin Principle. In line with our above remarks, we prove that the conditions of this principle are *sufficient* for weak first-order optimality and refer to rigorous proofs in the literature that the principle is a necessary condition for optimality; we comment here also on the notion of normality. We then remark that the two-point boundary-value problem that arises when terminal constraints are present is more involved than that in the unconstrained case; we briefly introduce a penalty function approach which circumvents this. Our results are then extended by allowing control constraints and by introducing strong *first-order* optimality. In particular, we show that Pontryagin's Principle in strong form is a sufficient condition for strong first-order optimality; again we refer to the literature for a proof that the principle is a necessary condition for optimality. Our next theorem allows the terminal time t_f to be unspecified. Finally, we obtain a sufficient condition for global optimality via a generalized Hamilton–Jacobi–Bellman equation. Throughout, examples are presented which illustrate and clarify the use of the theorems in control problems with terminal constraints.

4.2 Linear Dynamic System with General Performance Criterion and Terminal Equality Constraints

We derive a Pontryagin-type necessary condition for optimality for the case of a general performance criterion and both linear (Section 4.2.1) and nonlinear (Section 4.2.2) terminal equality constraint, but with linear system dynamics. These conditions, using elementary constructions, are adequate to deduce first-order necessary conditions for optimality.

4.2.1 Linear Dynamic System with Linear Terminal Equality Constraints

We return to the problem formulated in Section 3.2, where (3.1) is to be controlled to minimize the performance criterion (3.2). Now, however, we impose the restriction that, at the given final time t_f, the linear equality constraint

$$Dx(t_f) = 0 \tag{4.1}$$

be satisfied. Here, D is a $p \times n$ constant matrix. An important special case of the above formulation occurs when

$$\begin{aligned} L(x(t), u(t), t) &= u^T(t)u(t), \\ \phi(x(t_f)) &\equiv 0, \end{aligned} \tag{4.2}$$

and D is the $n \times n$ identity matrix. Then, the problem is one of steering the initial state of (3.1) from x_0 to the origin of the state space at time $t = t_f$ and minimizing the "energy"

$$J(u(\cdot); x_0) = \int_{t_0}^{t_f} u^T(t)u(t)\,dt. \tag{4.3}$$

Actually, it is easy to solve this particular problem directly, without appealing to the Pontryagin Principle derived later in this section. We need the following assumptions.

Assumption 4.2.1 *The control function $u(\cdot)$ is drawn from the set of piecewise continuous m-vector functions of t on the interval $[t_0, t_f]$ that meet the terminal constraints, $u(\cdot) \in \mathcal{U}_T$.*

Assumption 4.2.2 *The linear dynamic system* (3.1) *is controllable (see* [8]*) from $t = t_0$ to $t = t_f$, viz.,*

$$W(t_0, t_f) \triangleq \int_{t_0}^{t_f} \Phi(t_f, \tau) B(\tau) B^T(\tau) \Phi^T(t_f, \tau) d\tau \tag{4.4}$$

is positive definite, where

$$\frac{d}{dt}\Phi(t, \tau) = A(t)\Phi(t, \tau), \quad \Phi(\tau, \tau) = I. \tag{4.5}$$

With this assumption we can set

$$u^o(t) = -B^T(t)\Phi^T(t_f, t)W^{-1}(t_0, t_f)\Phi(t_f, t_0)x_0, \tag{4.6}$$

and, substituting this into (3.3), the solution of (3.1), we obtain

$$\begin{aligned} x^o(t_f) &= \Phi(t_f, t_0)x_0 - \left[\int_{t_0}^{t_f} \Phi(t_f, \tau)B(\tau)B^T(\tau)\Phi^T(t_f, \tau)d\tau\right] W^{-1}(t_0, t_f)\Phi(t_f, t_0)x_0 \\ &= \Phi(t_f, t_0)x_0 - \Phi(t_f, t_0)x_0 = 0 \end{aligned} \tag{4.7}$$

so that the control function $u^o(\cdot)$ steers x_0 to the origin of the state space at time $t = t_f$. Now suppose that $u(\cdot)$ is any other control function that steers x_0 to the origin at time t_f. Then, using (3.3) we have

$$0 = \Phi(t_f, t_0)x_0 + \int_{t_0}^{t_f} \Phi(t_f, \tau)B(\tau)u(\tau)d\tau \tag{4.8}$$

and

$$0 = \Phi(t_f, t_0)x_0 + \int_{t_0}^{t_f} \Phi(t_f, \tau)B(\tau)u^o(\tau)d\tau. \tag{4.9}$$

Subtracting these two equations yields

$$\int_{t_0}^{t_f} \Phi(t_f, \tau)B(\tau)[u(\tau) - u^o(\tau)]d\tau = 0, \tag{4.10}$$

and premultiplying by $2x_0\Phi^T(t_f, t_0)W^{-1}(t_0, t_f)$ yields

$$2\int_{t_0}^{t_f} u^{oT}(\tau)[u(\tau) - u^o(\tau)]d\tau = 0. \tag{4.11}$$

Subtracting the optimal control energy from any comparison control energy associated with $u(\cdot) \in \mathcal{U}_T$ and using (4.11), we have

$$\begin{aligned}
\Delta J &= \int_{t_0}^{t_f} u^T(t)u(t)dt - \int_{t_0}^{t_f} u^{oT}(t)u^o(t)dt \\
&= \int_{t_0}^{t_f} u^T(t)u(t)dt - \int_{t_0}^{t_f} u^{oT}(t)u^o(t)dt - 2\int_{t_0}^{t_f} u^{oT}(\tau)[u(\tau) - u^o(\tau)]d\tau \\
&= \int_{t_0}^{t_f} [u(t) - u^o(t)]^T[u(t) - u^o(t)]dt \geq 0,
\end{aligned} \tag{4.12}$$

which establishes that $u^o(\cdot)$ is minimizing, since for any $u \neq u^o$ the cost increases.

While the above treatment is quite adequate when (4.2) holds, it does not extend readily to the general case of (3.2) and (4.1). We therefore use the approach of Section 3.2.3, adjoining (3.1) by means of a continuously differentiable n-vector function of time $\lambda(\cdot)$ and (4.1) by means of a p-vector ν, as follows:

$$\begin{aligned}
\hat{J}(u(\cdot); \lambda(\cdot), \nu, x_0) \triangleq\ & J(u(\cdot); x_0) \\
&+ \int_{t_0}^{t_f} \lambda^T(t)[A(t)x(t) + B(t)u(t) - \dot{x}(t)]dt + \nu^T Dx(t_f).
\end{aligned} \tag{4.13}$$

Note that

$$\hat{J}(u(\cdot); \lambda(\cdot), \nu, x_0) = J(u(\cdot); x_0) \tag{4.14}$$

when (3.1) and (4.1) hold. Integrating by parts we obtain

$$\begin{aligned}\hat{J}(u(\cdot);\lambda(\cdot),\nu,x_0) &= J(u(\cdot);x_0)\\ &\quad+\int_{t_0}^{t_f}\left[\dot{\lambda}^T(t)x(t)+\lambda^T(t)A(t)x(t)+\lambda^T(t)B(t)u(t)\right]dt\\ &\quad+\lambda^T(t_0)x_0-\lambda^T(t_f)x(t_f)+\nu^T Dx(t_f).\end{aligned}\tag{4.15}$$

Let us suppose that there is a piecewise continuous control function $u^o(t)$ that minimizes (3.2) and causes

$$Dx^o(t_f)=0.\tag{4.16}$$

Next, evaluate the change in $\hat{J}$ brought about by changing $u^o(\cdot)$ to $u^o(\cdot)+\varepsilon\eta(\cdot)$. Note that the change in $\hat{J}$ is equal to the change in J if the perturbation $\varepsilon\eta(\cdot)$ is such that the perturbation in the trajectory at time $t=t_f$ satisfies

$$D[x^o(t_f)+\xi(t_f,\varepsilon)]=0,\tag{4.17}$$

i.e., $u^o(\cdot)+\varepsilon\eta(\cdot)\in\mathcal{U}_T$. Using (3.8), it follows that (4.17) holds only if

$$D\xi(t_f;\varepsilon)=\varepsilon Dz(t_f,\eta(\cdot))=0.\tag{4.18}$$

We return to this point later.

Referring to (3.14) and noting that (4.18) is relaxed below where it is adjoined to the perturbed cost by the Lagrange multiplier ν through the presence of $\nu^T Dx(t_f)$ in (4.15), we have

$$\begin{aligned}\Delta\hat{J} &= \hat{J}(u^o(\cdot)+\varepsilon\eta(\cdot);\lambda(\cdot),\nu,x_0)-\hat{J}(u^o(\cdot);\lambda(\cdot),\nu,x_0)\\ &= \int_{t_0}^{t_f}\Big[\varepsilon L_x(x^o(t),u^o(t),t)z(t;\eta(\cdot))+\varepsilon L_u(x^o(t),u^o(t),t)\eta(t)+\varepsilon\dot{\lambda}^T(t)z(t;\eta(\cdot))\\ &\qquad+\varepsilon\lambda^T(t)A(t)z(t;\eta(\cdot))+\varepsilon\lambda^T(t)B(t)\eta(t)+\mathcal{O}(t;\varepsilon)\Big]dt\\ &\quad-\varepsilon\lambda^T(t_f)z(t_f;\eta(\cdot))+\varepsilon\phi_x(x^o(t_f))z(t_f;\eta(\cdot))+\varepsilon\nu^T Dz(t_f;\eta(\cdot))+\mathcal{O}(\varepsilon).\end{aligned}\tag{4.19}$$

Now let us set

$$\begin{aligned} -\dot{\lambda}^T(t) &= L_x(x^o(t), u^o(t), t) + \lambda^T(t)A(t), \\ \lambda^T(t_f) &= \phi_x(x^o(t_f)) + \nu^T D. \end{aligned} \tag{4.20}$$

For fixed ν this is a legitimate choice for $\lambda(\cdot)$ as (4.20) is a linear ordinary differential equation in $\lambda(t)$ with piecewise continuous coefficients, having a unique solution. The right-hand side of (4.19) then becomes

$$\begin{aligned} \Delta\hat{J} &= \varepsilon \int_{t_0}^{t_f} \left[L_u(x^o(t), u^o(t), t) + \lambda^T(t)B(t)\right] \eta(t)dt \\ &\quad + \int_{t_0}^{t_f} \mathcal{O}(t;\varepsilon)dt + \mathcal{O}(\varepsilon). \end{aligned} \tag{4.21}$$

We set

$$\eta(t) = -\left[L_u(x^o(t), u^o(t), t) + \lambda^T(t)B(t)\right]^T, \tag{4.22}$$

which yields a piecewise continuous perturbation, and we now show that, under a certain assumption, a ν can be found such that $\eta(\cdot)$ given by (4.22) causes (4.18) to hold. We first introduce the following assumption.

Assumption 4.2.3 *The $p \times p$ matrix $\bar{W}(t_0, t_f)$ is positive definite where (see [8])*

$$\bar{W}(t_0, t_f) \triangleq \int_{t_0}^{t_f} D\Phi(t_f, \tau)B(\tau)B^T(\tau)\Phi^T(t_f, \tau)D^T d\tau = DW(t_0, t_f)D^T. \tag{4.23}$$

Note that this assumption is weaker than Assumption 4.2.2, being equivalent when $p = n$ and is called output controllability where the output is $y = Dx$.

From the linearity of (3.1) we have that $z(t;\eta(\cdot))$ satisfies the equation

$$\dot{z}(t;\eta(\cdot)) = A(t)z(t;\eta(\cdot)) + B(t)\eta(t), \quad z(t_0;\eta(\cdot)) = 0, \tag{4.24}$$

so that

$$z(t_f;\eta(\cdot)) = \int_{t_0}^{t_f} \Phi(t_f,\tau)B(\tau)\eta(\tau)d\tau, \tag{4.25}$$

and using (4.22)

$$z(t_f,\eta(\cdot)) = -\int_{t_0}^{t_f} \Phi(t_f,\tau)B(\tau)\left[L_u(x^o(\tau),u^o(\tau),\tau)+\lambda^T(\tau)B(\tau)\right]^T d\tau. \tag{4.26}$$

From (4.20) we have

$$\begin{aligned}\lambda(t) &= \Phi^T(t_f,t)\lambda(t_f) + \int_t^{t_f} \Phi^T(\tau,t)L_x^T(x^o(\tau),u^o(\tau),\tau)d\tau \\ &= \Phi^T(t_f,t)\phi_x^T(x^o(t_f)) + \Phi^T(t_f,t)D^T\nu \\ &\quad + \int_t^{t_f} \Phi^T(\bar\tau,t)L_x^T(x^o(\bar\tau),u^o(\bar\tau),\bar\tau)d\bar\tau. \end{aligned} \tag{4.27}$$

Premultiplying (4.26) by D and using (4.27), we obtain

$$\begin{aligned} &Dz(t_f;\eta(\cdot)) \\ &= -\int_{t_0}^{t_f} D\Phi(t_f,\tau)B(\tau)\Big[L_u^T(x^o(\tau),u^o(\tau),\tau)+B^T(\tau)\Phi^T(t_f,\tau)\phi_x^T(x^o(t_f)) \\ &\qquad + B^T(\tau)\int_\tau^{t_f} \Phi^T(\bar\tau,\tau)L_x^T(x^o(\bar\tau),u^o(\bar\tau),\bar\tau)d\bar\tau\Big]d\tau \\ &\quad - \int_{t_0}^{t_f} D\Phi(t_f,\tau)B(\tau)B^T(\tau)\Phi^T(t_f,\tau)D^T d\tau\nu. \end{aligned} \tag{4.28}$$

Setting the left-hand side of (4.28) equal to zero we can, in view of Assumption 4.2.3, uniquely solve for ν in terms of the remaining (all known) quantities in (4.28). Consequently, we have proved that there exists a ν, independent of ε, such that $\eta(\cdot)$ given by (4.22) causes (4.18) to be satisfied. With this choice of $\eta(\cdot)$, (4.14) holds, and we have that the change in $\hat{J}$ is

$$\Delta\hat{J} = -\varepsilon\int_{t_0}^{t_f} \|L_u(x^o(t),u^o(t),t)+\lambda^T(t)B(t)\|^2 dt + \int_{t_0}^{t_f} \mathcal{O}(t;\varepsilon)dt + \mathcal{O}(\varepsilon) \geq 0. \tag{4.29}$$

Using the usual limiting argument that $\mathcal{O}(t;\varepsilon)/\varepsilon \to 0$ and $\mathcal{O}(\varepsilon)/\varepsilon \to 0$ as $\varepsilon \to 0$, then since the first-order term dominates for optimality as $\varepsilon \to 0$,

$$\int_{t_0}^{t_f} \|L_u(x^o(t), u^o(t), t) + \lambda^T(t)B(t)\|^2 dt \geq 0. \tag{4.30}$$

It then follows that a necessary condition for $u^o(\cdot)$ to minimize J is that there exists a p-vector ν such that

$$L_u(x^o(t), u^o(t), t) + \lambda^T(t)B(t) = 0 \quad \forall\, t \text{ in } [t_0, t_f], \tag{4.31}$$

where

$$-\dot{\lambda}^T(t) = L_x(x^o(t), u^o(t), t) + \lambda^T(t)A(t), \quad \lambda^T(t_f) = \phi_x(x^o(t_f)) + \nu^T D. \tag{4.32}$$

As in Section 3.2.6, the necessary condition derived above can be restated in terms of the Hamiltonian to yield a weak version of Pontryagin's Principle for the optimal control problem formulated with linear dynamics and linear terminal constrants.

Theorem 4.2.1 *Suppose that* $u^o(\cdot)$ *minimizes the performance criterion* (3.2) *subject to the dynamic system* (3.1) *and the terminal constraint* (4.1) *and that Assumptions* 3.2.1, 3.2.3, 3.2.4, 4.2.1, *and* 4.2.3 *hold. Suppose further that* $H(x, u, \lambda, t)$ *is defined according to* (3.23). *Then, there exists a p-vector* ν *such that the partial derivative of the Hamiltonian with respect to the control* $u(t)$ *is zero when evaluated at the optimal state and control* $x^o(t)$, $u^o(t)$, *viz.,*

$$H_u(x^o(t), u^o(t), \lambda(t), t) = 0 \quad \forall\, t \text{ in } [t_0, t_f], \tag{4.33}$$

where

$$-\dot{\lambda}^T(t) = H_x(x^o(t), u^o(t), \lambda(t), t), \quad \lambda^T(t_f) = \phi_x(x^o(t_f)) + \nu^T D. \tag{4.34}$$

4.2.2 Pontryagin Necessary Condition: Special Case

We return to the special case specified by (4.2). From (4.7) it follows that $u^o(\cdot)$ given by (4.6) steers x_0 to the origin at time t_f, and from (4.12) we concluded that $u^o(\cdot)$ is in fact minimizing. We now show that $u^o(\cdot)$ satisfies Pontryagin's necessary condition. Indeed, by substituting (4.6) into (4.28), ν becomes with $D = I$

$$\nu = 2W^{-1}(t_0, t_f)\Phi(t_f, t_0)x_0. \tag{4.35}$$

Also, from (4.27) and with $D = I$,

$$\lambda(t) = 2\Phi^T(t_f, t)W^{-1}(t_0, t_f)\Phi(t_f, t_0)x_0. \tag{4.36}$$

Finally,

$$H_u(x^o(t), u^o(t), \lambda, t) = 2u^{oT}(t) + 2B^T(t)\Phi^T(t_f, t)W^{-1}(t_0, t_f)\Phi(t_f, t_0)x_0, \tag{4.37}$$

which, by (4.6), is zero for all t in $[t_0, t_f]$.

4.2.3 Linear Dynamics with Nonlinear Terminal Equality Constraints

We now turn to a more general terminal constraint than (4.1), viz.,

$$\psi(x(t_f)) = 0, \tag{4.38}$$

where $\psi(\cdot)$ is a p-dimensional vector function of its n-dimensional argument, which satisfies the following assumption.

Assumption 4.2.4 *The p-dimensional function $\psi(\cdot)$ is once continuously differentiable in x.*

As a consequence of Assumption 4.2.4 we can write

$$\psi(x^o(t_f) + \varepsilon z(t_f; \eta(\cdot))) = \psi(x^o(t_f)) + \varepsilon\psi_x(x^o(t_f))z(t_f; \eta(\cdot)) + \mathcal{O}(\varepsilon). \tag{4.39}$$

The change in J, caused by a perturbation $\varepsilon\eta(\cdot)$ in the control function which is so chosen that

$$\psi(x^o(t_f) + \xi(t_f;\varepsilon)) = 0, \tag{4.40}$$

is therefore given by (4.19) with D replaced by $\psi_x(x^o(t_f))$. We can then proceed to set

$$\begin{aligned} -\dot{\lambda}^T(t) &= L_x(x^o(t), u^o(t), t) + \lambda^T(t)A(t), \\ \lambda^T(t_f) &= \phi_x(x^o(t_f)) + \nu^T\psi_x(x^o(t_f)) \end{aligned} \tag{4.41}$$

to yield, as before, the change in J as given by (4.21).

If we then specify $\eta(\cdot)$ by (4.22), we cannot directly show (i.e., by solving for ν using elementary mathematics) that there exists a ν such that, when $u^o(\cdot) + \varepsilon\eta(\cdot)$ is applied to (3.1), (4.40) is satisfied; deeper mathematics[6] is required. However, it follows directly from (4.28) that we could calculate a ν such that

$$\psi_x(x^o(t_f))z(t_f;\eta(\cdot)) = 0 \tag{4.42}$$

by replacing in (4.28) D by $\psi_x(x^o(t_f))$. In other words, we could show that there exists a ν such that $u^o(\cdot) + \varepsilon\eta(\cdot)$ satisfies (4.40) to first-order in ε. If we then replace the requirement (4.40) by (4.42) we would arrive at Theorem 4.2.1 with D replaced by $\psi_x(x^o(t_f))$. Although this is in fact the correct Pontryagin necessary condition for this problem, we would have arrived at it nonrigorously, viz., by replacing (4.40) with (4.42), without rigorously justifying the satisfaction of the equality constraint to only first order.

[6]A form of the Implicit Function Theorem associated with constructing controls that meet nonlinear constraints exactly [38] is needed. [28] gives some insight into the issues for nonlinear programming problems.

In view of the above, we prefer to show that Pontryagin's Principle is a sufficient condition for "first-order optimality"; this permits a rigorous treatment. Therefore, as stated in Section 4.1, the emphasis of this book, as far as proofs are concerned, swings now to developing sufficient conditions for first-order optimality. Reference is made to rigorous proofs, where these exist, that the conditions are also necessary for optimality.

As outlined above, it is the nonlinearity of the terminal constraint that makes our straightforward approach inadequate for the derivation of Pontryagin's Principle. Once a nonlinear terminal constraint is present and we have opted for showing that Pontryagin's Principle is sufficient for first-order optimality, nothing is gained by restricting attention to the linear dynamic system (3.1). Consequently we treat the problem of minimizing (3.2) subject to (3.41) and (4.38).

4.3 Weak First-Order Optimality with Nonlinear Dynamics and Terminal Constraints

We allow weak perturbations in the control of the form $u^o(\cdot) + \varepsilon\eta(\cdot)$, where $\eta(\cdot)$ is a piecewise continuous m-vector function on $[t_0, t_f]$. We prove that the conditions for the weak form of the Pontryagin Principle are sufficient for weak first-order optimality, which is defined as follows.

Definition 4.3.1 *J, given in* (3.2)*, is weakly first-order optimal subject to* (3.41) *and* (4.38) *at $u^o(\cdot)$ if the first-order change in J, caused by any weak perturbation $\varepsilon\eta(\cdot)$ of $u^o(\cdot)$, which maintains* (4.38) *to be satisfied or $u^o(\cdot) + \varepsilon\eta(\cdot) \in \mathcal{U}_T$, is nonnegative.*

Definition 4.3.2 *J, given in* (3.2)*, is weakly locally optimal subject to* (3.41) *and* (4.38) *at $u^o(\cdot)$ if there exists an $\bar{\varepsilon} > 0$ such that the change in J is nonnegative for* *all perturbations* $\bar{\eta}(\cdot)$ *of* $u^o(\cdot)$ *which maintain* (4.38) *and which satisfy* $\|\bar{\eta}(\cdot)\| \le \bar{\varepsilon}$ *for all* t *in* $[t_0, t_f]$, *or*

$$\Delta \hat{J} \ge 0 \quad \forall \quad u^o(\cdot) + \bar{\eta}(\cdot) \in \mathcal{U}_T. \tag{4.43}$$

4.3.1 Sufficient Condition for Weakly First-Order Optimality

Weakly first-order optimal does not usually imply weakly locally optimal. Weakly first-order optimal merely implies that if J is not weakly locally optimal at $u^o(\cdot)$, this must be owing to higher-order, and not first-order, effects.

Following the approach in (4.13) and (4.15), we adjoin the system equation (3.41) and the terminal constraint (4.38) to J and integrate by parts to obtain

$$\begin{aligned} \hat{J}(u(\cdot); \lambda(\cdot), \nu, x_0) &= J(u(\cdot); x_0) + \int_{t_0}^{t_f} \left[\dot{\lambda}^T(t) x(t) + \lambda^T(t) f(x(t), u(t), t)\right] dt \\ &\quad + \lambda^T(t_0) x_0 - \lambda^T(t_f) x(t_f) + \nu^T \psi(x(t_f)), \end{aligned} \tag{4.44}$$

and it follows that

$$\hat{J}(u(\cdot); \lambda(\cdot), \nu, x_0) = J(u(\cdot); x_0) \tag{4.45}$$

whenever (3.41) and (4.38) are satisfied.

We now evaluate the change in $\hat{J}$ (i.e., in J) brought about by changing $u^o(\cdot)$ to $u^o(\cdot) + \varepsilon\eta(\cdot)$ keeping (4.38) satisfied. Using (3.52) this change is

$$\begin{aligned} \Delta \hat{J} &= \hat{J}(u^o(\cdot) + \varepsilon\eta(\cdot); \lambda(\cdot), \nu, x_0) - \hat{J}(u^o(\cdot); \lambda(\cdot), \nu, x_0) \\ &= \int_{t_0}^{t_f} [\varepsilon H_x(x^o(t), u^o(t), \lambda(t), t) z(t; \eta(\cdot)) + \varepsilon \dot{\lambda}^T(t) z(t; \eta(\cdot)) \\ &\quad + \varepsilon H_u(x^o(t), u^o(t), \lambda(t), t)\eta(t) + \mathcal{O}(t; \varepsilon)]\, dt - \varepsilon \lambda^T(t_f) z(t_f; \eta(\cdot)) \\ &\quad + \varepsilon \phi_x(x^o(t_f)) z(t_f; \eta(\cdot)) + \varepsilon \nu^T \psi_x(x^o(t_f)) z(t_f; \eta(\cdot)) + \mathcal{O}(\varepsilon). \end{aligned} \tag{4.46}$$

Now, if we set

$$
\begin{aligned}
-\dot{\lambda}^T(t) &= H_x(x^o(t), u^o(t), \lambda(t), t), \\
\lambda^T(t_f) &= \phi_x(x^o(t_f)) + \nu^T \psi_x(x^o(t_f)),
\end{aligned}
\tag{4.47}
$$

we see that a *sufficient* condition for the change in J to be nonnegative (in fact, zero) to first order in ε is that

$$
H_u(x^o(t), u^o(t), \lambda(t), t) = 0 \quad \forall\, t \text{ in } [t_0, t_f]. \tag{4.48}
$$

Remark 4.3.1 *Although the derivation of the change in $\hat{J}$ is similar to that given in Chapter 3, because no terminal constraints had to be maintained in the presence of arbitrary variations, we could conclude that the first-order conditions were necessary for weakly local optimality. Here, because we have derived a sufficient condition for weak first-order optimality, it was not necessary to actually construct a perturbation $\varepsilon\eta(\cdot)$ which maintains satisfaction of* (4.38)*. All that was necessary was to note that if* (4.48) *holds, then J is nonnegative to first order for any weak perturbation that maintains satisfaction of* (4.38).

Before we state our results, we introduce the notion of normality.

Remark 4.3.2 (Normality) *In the derivations of Pontryagin's Principle as a necessary condition for optimality, to which we refer, the Hamiltonian H is defined as*

$$
H(x, u, \lambda, \lambda_0, t) = \lambda_0 L(x, u, t) + \lambda^T f(x, u, t), \tag{4.49}
$$

where

$$
\lambda_0 \geq 0, \tag{4.50}
$$

and whenever ϕ or one of its derivatives appears in a formula, it is premultiplied by λ_0. The statement of the principle is then that if $u^o(\cdot)$ minimizes J subject to the *dynamic system, control and terminal constraints, then there exist* $\lambda_0 \geq 0$ *and* ν *not all zero, such that certain conditions hold.*

If on an optimal path the principle can be satisfied upon setting $\lambda_0 = 1$ *(any positive* λ_0 *can be normalized to unity because of the homogeneity of the expressions involved), the problem is referred to as "normal." Throughout this book we assume that when we refer to statements and proofs of Pontryagin's Principle as a necessary condition for optimality, the problem is normal; consequently,* λ_0 *does not appear in any of our theorems.*

It is evident from the proofs of the necessary conditions in Chapter 3 *that every free end-point control problem is normal and the control problem with linear dynamics and linear terminal constraints is normal if Assumption* 4.2.3 *is satisfied.*

Theorem 4.3.1 *Suppose that Assumptions* 3.2.3, 3.2.4, 3.3.1, 3.3.2, 4.2.1, *and* 4.2.4 *are satisfied. Suppose further that* $u^o(\cdot) \in \mathcal{U}_T$ *minimizes the performance criterion* (3.2) *subject to* (3.41) *and* (4.38). *Then there exists a p-vector* ν *such that*

$$H_u(x^o(t), u^o(t), \lambda(t), t) = 0 \quad \forall\ t \text{ in } [t_0, t_f], \tag{4.51}$$

where

$$\begin{aligned} -\dot{\lambda}^T(t) &= H_x(x^o(t), u^o(t), \lambda(t), t), &(4.52)\\ \lambda^T(t_f) &= \phi_x(x^o(t_f)) + \nu^T \psi_x(x^o(t_f)), &(4.53) \end{aligned}$$

and

$$H(x, u, \lambda, t) \triangleq L(x, u, t) + \lambda^T f(x, u, t). \tag{4.54}$$

Moreover, the above condition is sufficient for J to be weakly first-order optimal at $u^o(\cdot)$.

Proof: A rigorous proof that Pontryagin's Principle is a necessary condition for optimality, as stated in Section 4.2.3, is beyond the scope of this book. Rigorous proofs are available in [44, 10, 25, 38]. Upon assuming that the problem is normal, the above conditions result. The second part of the theorem is proved in Section 4.3.1. ∎

Example 4.3.1 (Rocket launch example) *Let us consider the problem of maximizing the terminal horizontal velocity component of a rocket in a specified time* $[t_0, t_f]$ *subject to a specified terminal altitude and a specified terminal vertical velocity component. This launch is depicted in Figure* 4.1.

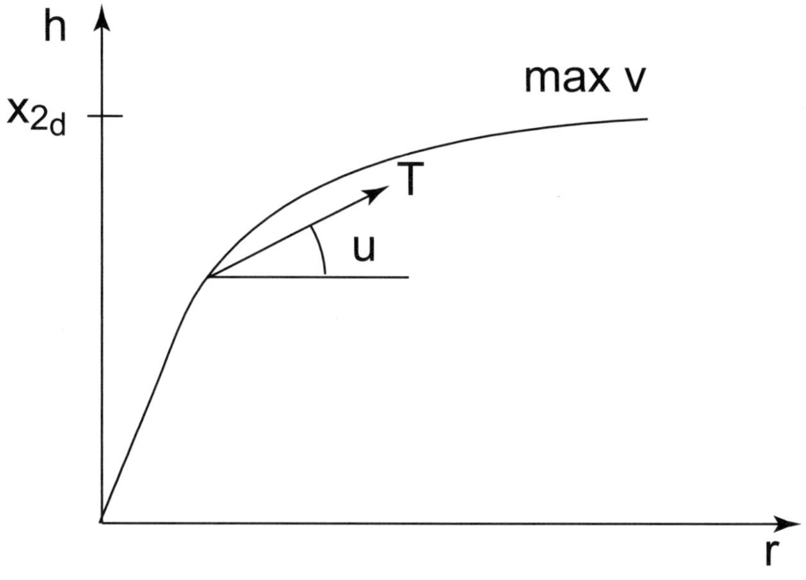

Figure 4.1: Rocket launch example.

A simplified mathematical model of the vehicle is

$$\begin{aligned}
\dot{x}_1(t) &= \dot{r} = x_3(t), \\
\dot{x}_2(t) &= \dot{h} = x_4(t), \\
\dot{x}_3(t) &= \dot{v} = T\cos u(t), \\
\dot{x}_4(t) &= \dot{w} = T\sin u(t) - g,
\end{aligned} \tag{4.55}$$

where $x_1(t)$ *is the horizontal component of the position of the vehicle, at time* t, $t_0 \le t \le t_f$, $x_2(t)$ *is the altitude or vertical component of its position,* $x_3(t)$ *is its horizontal component of velocity,* $x_4(t)$ *is its vertical component of velocity,* $u(t)$ *is the inclination of the rocket motor's thrust vector to the horizontal,* g *is the (constant) gravitational acceleration, and* T *is the constant specific thrust of the rocket motor. We suppose the following initial conditions for the rocket at time* $t = t_0$*:*

$$
\begin{aligned}
x_1(t_0) &= x_{10}, \\
x_2(t_0) &= x_{20}, \\
x_3(t_0) &= 0, \\
x_4(t_0) &= 0.
\end{aligned}
\tag{4.56}
$$

The problem, then, is to determine $u(\cdot)$ *in the interval* $[t_0, t_f]$ *to minimize*

$$
J(u(\cdot); x_0) = -x_3(t_f) \tag{4.57}
$$

subject to the terminal constraints

$$
x_2(t_f) = x_{2d}, \quad x_4(t_f) = x_{4d}, \tag{4.58}
$$

where x_{2d} *and* x_{4d} *are, respectively, the desired altitude and vertical velocity. From* (4.54) *we have*

$$
H(x, u, \lambda, t) = \lambda_1 x_3 + \lambda_2 x_4 + \lambda_3 T \cos u + \lambda_4 T \sin u - \lambda_4 g, \tag{4.59}
$$

and from (4.52) *and* (4.53)

$$
\begin{aligned}
-\dot{\lambda}_1(t) &= 0, & \lambda_1(t_f) &= 0, \\
-\dot{\lambda}_2(t) &= 0, & \lambda_2(t_f) &= \nu_2, \\
-\dot{\lambda}_3(t) &= \lambda_1(t), & \lambda_3(t_f) &= -1, \\
-\dot{\lambda}_4(t) &= \lambda_2(t), & \lambda_4(t_f) &= \nu_4.
\end{aligned}
\tag{4.60}
$$

From (4.60) *we conclude that*

$$\begin{aligned} \lambda_1(t) &= 0, \\ \lambda_2(t) &= \nu_2, \\ \lambda_3(t) &= -1, \\ \lambda_4(t) &= \nu_4 + (t_f - t)\nu_2 \end{aligned} \tag{4.61}$$

for t *in* $[t_0, t_f]$. *From* (4.59) *we have*

$$H_u(x, u, \lambda, t) = -\lambda_3 T \sin u + \lambda_4 T \cos u. \tag{4.62}$$

Therefore, using (4.62), (4.51) *is satisfied if and only if*

$$\begin{aligned} u^o(t) &= \arctan\left[\frac{\lambda_4(t)}{\lambda_3(t)}\right] \\ &= \arctan\left[-\nu_4 - (t_f - t)\nu_2\right]. \end{aligned} \tag{4.63}$$

Equation (4.63) *gives the form of the control as a function of* t, *which satisfies Theorem* 4.3.1. *Naturally, the free parameters* ν_2 *and* ν_4 *have to be chosen so that when* (4.63) *is applied to* (4.55), *the terminal equality constraints* (4.58) *are satisfied. What is important to note here is that the form of* $u(\cdot)$ *has been determined, which is necessary for the minimization of* (4.57). *However, the optimal solution needs to be resolved numerically.*

Numerical Optimization with Terminal Constraints: The Penalty Function Approach

When terminal constraints (4.38) are present, the two-point boundary-value problem outlined in Section 3.3.4 is more complex because of the presence of the unknown parameters ν in the expression for $\lambda(t)$. Whereas in Section 4.2.2 it was possible to find an explicit expression determining ν, in general this is not possible for nonlinear constraints. We conclude that, given a candidate pair $(\tilde{u}(\cdot),\ \tilde{x}(\cdot))$ which satisfies (4.38), it is not possible to check directly whether (4.51) is satisfied, as was the case in Section 3.3.4; rather, one would have to determine, numerically, whether there exists a ν which, when used in (4.52) and (4.54), yields (4.51).

Similar remarks hold when constructing a control function $\tilde{u}(\cdot)$ which satisfies (4.38) and Pontryagin's Principle. Here, one would have to resort to numerical techniques to determine both $\tilde{u}(\cdot)$ and ν. As indicated for a special case in Section 4.2.2, the class of linear quadratic optimal control problems allows an explicit solution; this is discussed further in Chapter 5.

It turns out that engineers quite often sidestep the difficulty of determining the additional parameters ν by converting the control problem with a terminal constraint (4.38) into an approximately equivalent unconstrained one by adding the term $\rho\psi^T(x(t_f))\psi(x(t_f))$ to J to form a new performance criterion,

$$\bar{J}(u(\cdot);\rho,x_0) = J(u(\cdot);x_0) + \rho\psi^T(x(t_f))\psi(x(t_f)), \tag{4.64}$$

where $\rho > 0$. Minimization of (4.64) with respect to $u(\cdot)$, for ρ sufficiently large, causes (4.38) to be approximately satisfied. Indeed, it can be shown that under fairly weak assumptions, the minimum of (4.64) with respect to $u(\cdot)$ tends to the minimum of (3.2) subject to (4.38), as $\rho \to \infty$; this is the so-called penalty function method of satisfying terminal equality constraints [11].

Steepest Descent Approach to Terminally Constrained Optimization Problems

The following steepest descent algorithm [11] is a method for finding a constrained minimum. Choose a nominal control $u_N(t)$ and let $x_N(t)$ be the resulting path. This path satisfies the nonlinear dynamic equation

$$\dot{x}_N(t) = f(x_N(t), u_N(t), t) \Rightarrow x_N(\cdot). \tag{4.65}$$

This path may be nonoptimal and the terminal constraint may not be satisfied, $\psi(x_N(t_f)) \neq 0$.

Consider perturbations in $u_N(t)$ as

$$u_{N+1}(\cdot) = u_N(\cdot) + \delta u(\cdot), \tag{4.66}$$

where $\delta u(\cdot) = \varepsilon\eta(\cdot)$. Then, the first-order term in the Taylor expansion of the dynamics is

$$\delta\dot{x}(t) = f_x(x_N(t), u_N(t), t)\delta x(t) + f_u(x_N(t), u_N(t), t)\delta u(t), \tag{4.67}$$

where $\delta x(t) = \varepsilon z(t; \eta(\cdot))$. The objective is to predict how the control perturbation $\delta u(t)$ will affect the cost criterion and the terminal constraint. To construct these predictions, consider the influence function $\lambda^{\psi}(t) \in \mathbb{R}^{n\times p}$ associated with the terminal constraint functions as

$$\dot{\lambda}^{\psi}(t) = -f_x^T(x_N(t), u_N(t), t)\lambda^{\psi}(t), \quad \lambda^{\psi}(t_f) = \psi_x^T(t_f); \tag{4.68}$$

then, the change in the terminal constraint is

$$\delta\psi(t_f) = \psi_x(t_f)\delta x(t_f) = \lambda^{\psi^T}(t_0)\delta x(t_0) + \int_{t_0}^{t_f} \lambda^{\psi^T}(t) f_u(x_N(t), u_N(t), t)\delta u(t)\, dt, \tag{4.69}$$

where for $x(t_0)$, fixed $\delta x(t_0) = 0$. Similarly, the change in the performance index is

$$\begin{aligned} &J(u_{N+1}(\cdot), x_o) - J(u_N(\cdot), x_o) = \delta J \\ &= \phi_x \delta x(t_f) + \int_{t_0}^{t_f} (L_x(x_N(t), u_N(t), t)\delta x(t) + L_u(x_N(t), u_N(t), t)\delta u(t))dt \qquad (4.70) \\ &= \lambda^{\phi}(t_0)^T \delta x(t_0) + \int_{t_0}^{t_f} (L_u(x_N(t), u_N(t), t) \\ &\quad + \lambda^{\phi}(t)^T f_u(x_N(t), u_N(t), t))\delta u(t)\, dt, \qquad (4.71) \end{aligned}$$

where

$$\dot{\lambda}^{\phi}(t) = -f_x^T(x_N(t), u_N(t), t)\lambda^{\phi}(t) - L_x^T(x_N(t), u_N(t), t), \quad \lambda^{\phi}(t_f) = \phi_x(t_f)^T, \tag{4.72}$$

and the zero term

$$\int_{t_0}^{t_f} \frac{d}{dt}\left(\lambda^{\phi^T}(t)\delta x(t)\right) dt - \int_{t_0}^{t_f} (-L_x(x_N(t), u_N(t), t)\delta x(t) + \lambda^{\phi^T}(t) f_u(x_N(t), u_N(t), t)\delta u(t))dt = 0 \tag{4.73}$$

was subtracted from (4.70) to obtain, after some manipulations, (4.71). To make an improvement in δJ and decrease the constraint violation, adjoin $\delta\psi(t_f)$ to δJ with a Lagrange multiplier $\nu \in \mathbb{R}^p$ as

$$\begin{aligned}\delta J + \nu^T \delta\psi(t_f) &= \left(\lambda^{\phi^T}(t_0) + \nu^T \lambda^{\psi^T}(t_0)\right) \delta x(t_0) \\ &\quad + \int_{t_0}^{t_f} \left[\left(\lambda^{\phi^T}(t) + \nu^T \lambda^{\psi^T}(t)\right)(t) f_u(x_N(t), u_N(t), t) \right. \\ &\qquad \left. + L_u(x_N(t), u_N(t), t)\right] \delta u(t)\, dt, \end{aligned} \tag{4.74}$$

where ν is chosen so that a desired change in $\delta\psi(t_f)$ is met. Choose

$$\begin{aligned}&\delta u(t) \\ &= -\epsilon \left[\left(\lambda^{\phi^T}(t) + \nu^T \lambda^{\psi^T}(t)\right) f_u(x_N(t), u_N(t), t) + L_u(x_N(t), u_N(t), t)\right]^T,\end{aligned} \tag{4.75}$$

$$\begin{aligned}&\psi_x(x_N(t_f))\delta x(t_f) \\ &= -\epsilon \int_{t_0}^{t_f} \lambda^{\psi^T}(t) f_u(x_N(t), u_N(t), t) f_u^T(x_N(t), u_N(t), t) \left(\lambda^{\phi^T}(t) + \nu^T \lambda^{\psi^T}(t)\right)^T dt \\ &\quad -\epsilon \int_{t_0}^{t_f} \lambda^{\psi}(t)^T f_u(x_N(t), u_N(t), t) L_u^T(x_N(t), u_N(t), t) dt,\end{aligned} \tag{4.76}$$

then

$$\begin{aligned}&\psi_x(x_N(t_f))\delta x(t_f) + \epsilon \int_{t_0}^{t_f} \lambda^{\psi^T}(t) f_u(x_N(t), u_N(t), t) L_u^T(x_N(t), u_N(t), t) dt \\ &\quad + \epsilon \int_{t_0}^{t_f} \lambda^{\psi^T}(t) f_u(x_N(t), u_N(t), t) f_u^T(x_N(t), u_N(t), t) \lambda^{\phi}(t) dt \\ &= -\epsilon \int_{t_0}^{t_f} \lambda^{\psi^T}(t) f_u(x_N(t), u_N(t), t) f_u^T(x_N(t), u_N(t), t) \lambda^{\psi}(t) dt\ \nu.\end{aligned} \tag{4.77}$$

Solve for ν as

$$\begin{aligned}\nu &= -\left[\int_{t_0}^{t_f} \lambda^{\psi^T}(t) f_u(x_N(t), u_N(t), t) f_u^T(x_N(t), u_N(t), t)\lambda^{\psi}(t)dt\right]^{-1} \\ &\quad \times \left[\psi_x(x_N(t_f))\delta x(t_f)/\epsilon + \int_{t_0}^{t_f} \lambda^{\psi^T}(t) f_u(x_N(t), u_N(t), t) L_u^T(x_N(t), u_N(t), t)dt \right. \\ &\quad \left. + \int_{t_0}^{t_f} \lambda^{\psi^T}(t) f_u(x_N(t), u_N(t), t) f_u^T(x_N(t), u_N(t), t)\lambda^{\phi}(t)dt\right], \end{aligned} \tag{4.78}$$

where the inverse exists from Assumption 4.2.3 and $D = \psi_x(x_N(t_f))$.

If $\delta\psi(x_N(t_f)) = \psi_x \delta x(t_f) = 0$ and $\delta x(t_0) = 0$, then note from (4.69) that the functions $\lambda^{\psi^T}(t) f_u(x_N(t), u_N(t), t)$ and $(\lambda^{\phi^T}(t) + \nu^T \lambda^{\psi^T}(t)) f_u(x_N(t), u_N(t), t) + L_u(x_N(t), u_N(t), t)$ are orthogonal. In any case,

$$\begin{aligned}\delta \hat{J} = \delta J + \nu \delta\psi(t_f) &= -\epsilon \int_{t_0}^{t_f} \left\| \left(\lambda^{\phi^T}(t) + \nu^T \lambda^{\psi^T}(t)\right) f_u(x_N(t), u_N(t), t) \right. \\ &\qquad \left. + L_u(x_N(t), u_N(t), t) \right\|^2 dt, \end{aligned} \tag{4.79}$$

and therefore the cost becomes smaller on each iteration.

The above algorithm is summarized in the following steps:

1. Choose the nominal control $u_N(t)$ over the interval $t \in [t_0,\, t_f]$.

2. Integrate (4.65) forward from the initial conditions to the terminal time t_f to obtain the nominal state $x_N(t)$. Store the values over the trajectory.

3. Integrate (4.68) and (4.72) backward from its terminal condition to the initial time t_0 and store the values over the trajectory.

4. Choose the desired change in the terminal constraint $\delta\psi(t_f)$ and the desired change in the cost through the choice of ϵ. These choices are made small enough to retain the assumed linearity but large enough for the algorithm to converge quickly.

5. Compute ν from (4.78). The integrals in (4.78) can be computed along with the influence functions obtained from (4.68) and (4.72).

6. Compute the perturbation $\delta u(t)$ from (4.75) and form a new nominal control as given in (4.66) and repeat step 2. Check to see if the actual change in the constraints and cost are close to the predicted values of the perturbed constraint (4.69) and perturbed cost (4.71) based on the assumed linear perturbation theory. This is where the choices of ϵ and $\delta\psi(t_f)$ are checked to determine if they are made small enough to retain the assumed linearity but large enough for the algorithm to converge quickly.

4.4 Strong First-Order Optimality

It is clear that if

$$H(x^o(t), u^o(t), \lambda(t), t) \le H(x^o(t), u(t), \lambda(t), t) \quad \forall\, u(t) \in \mathcal{U}_T, \tag{4.80}$$

then (4.51) is satisfied. Consequently, (4.51)–(4.54) with (4.51) replaced by (4.80) are also sufficient conditions for J to be weakly first-order optimal at $u^o(\cdot)$. It is clear also that (4.80) ensures that J is optimal to first order for a perturbation $\eta(\cdot;\varepsilon)$ which is made up of a weak perturbation $\varepsilon\eta(\cdot)$ and a strong perturbation (Section 3.4) which maintains (4.38), i.e., a perturbation of the form

$$\eta(t;\varepsilon) = \begin{cases} \eta(t), & \bar{t} \le t \le \bar{t} + \varepsilon, \\ \varepsilon\eta(t) & \text{in } [t_0, \bar{t}) \text{ and } (\bar{t} + \varepsilon, t_f]. \end{cases} \tag{4.81}$$

However, there are more elaborate strong perturbations than (4.81). Indeed, we shall permit the larger class of strong perturbations $\eta(\cdot;\varepsilon)$ given by

$$\eta(t;\varepsilon) = \begin{cases} \eta(t), & t_i \le t \le t_i + \varepsilon\delta_i, \quad i = 1, \dots, N, \\ \varepsilon\eta(t), & t \in I \triangleq \left\{ [t_0, t_f] - \bigcup_{i=1}^{n} [t_i, t_i + \varepsilon\delta_i] \right\}, \end{cases} \tag{4.82}$$

where $\eta(\cdot)$ is continuous on the nonoverlapping intervals $[t_i, t_i+\varepsilon\delta_i]$, $i = 1, \dots, N$, and is piecewise continuous on the remaining subintervals of $[t_0, t_f]$. This class is much larger than (4.81), since N, $\delta_i > 0$, and $\eta(\cdot)$ are arbitrary. This richer class is used to satisfy the terminal constraints while optimizing the cost criterion. The intervals $[t_i, t_i+\varepsilon\delta_i]$, although nonoverlapping, may be contiguous. We then have the following definitions.

Definition 4.4.1 *J of* (3.2) *is strongly first-order optimal subject to* (3.41) *and* (4.38) *at $u^o(\cdot)$ if the first-order change in J, caused by any strong perturbation $\eta(\cdot;\varepsilon)$ of the form* (4.82)*, which maintains* (4.38)*, is nonnegative.*

Definition 4.4.2 *J of* (3.2) *is strongly locally optimal subject to* (3.41) *and* (4.38) *at $u^o(\cdot)$ if there exists an $\bar{\varepsilon} > 0$ such that the change in J is nonnegative for all perturbations $\eta(\cdot;\varepsilon)$ of $u^o(\cdot)$ which maintain* (4.38)*, and which satisfy $\|\xi(t;\eta(\cdot;\varepsilon))\| \le \bar{\varepsilon}$, for all t in $[t_0, t_f]$.*

Note that, similar to the weak first-order optimality, strong first-order optimality does not usually imply strong local optimality. Strong first-order optimality merely implies that if J is not strongly locally optimal at $u^o(\cdot)$, this must be because of higher-order, and not first-order, effects.

We have the expression

$$\begin{aligned}\hat{J}(u(\cdot);\lambda(\cdot),\nu,x_0) &= \int_{t_0}^{t_f}\left[H(x(t),u(t),\lambda(t),t) + \dot{\lambda}^T(t)x(t)\right]dt \\ &\quad + \phi(x(t_f)) + \nu^T\psi(x(t_f)) + \lambda^T(t_0)x_0 - \lambda^T(t_f)x(t_f), \quad (4.83)\end{aligned}$$

which is equal to $J(u(\cdot);x_0)$ whenever (3.41) and (4.38) are satisfied. We now evaluate the change in $\hat{J}$ (i.e., in J) caused by changing $u^o(\cdot)$ to $u^o(\cdot) + \eta(\cdot;\varepsilon)$ while keeping

(4.38) satisfied.

$$
\begin{aligned}
\Delta \hat{J} &= \hat{J}(u^o(\cdot)+\eta(\cdot;\varepsilon);\lambda(\cdot),\nu,x_0) - \hat{J}(u^o(\cdot);\lambda(\cdot),\nu,x_0) \\
&= \int_{t_0}^{t_f} \Big[H(x^o(t)+\xi(t;\eta(\cdot;\varepsilon)), u^o(t)+\eta(t;\varepsilon), \lambda(t), t) \\
&\qquad - H(x^o(t), u^o(t), \lambda(t), t) + \dot{\lambda}^T(t)\xi(t;\eta(\cdot;\varepsilon)) \Big] dt \\
&\quad + \phi(x^o(t_f)+\xi(t_f;\varepsilon)) - \phi(x^o(t_f)) \qquad (4.84) \\
&\quad + \nu^T\psi(x^o(t_f)+\xi(t_f;\eta(\cdot;\varepsilon)) - \nu^T\psi(x^o(t_f)) - \lambda^T(t_f)\xi(t_f;\eta(\cdot;\varepsilon)).
\end{aligned}
$$

Along the lines of Sections 3.3 and 3.4, one can show that

$$
\xi(t;\eta(\cdot;\varepsilon)) = \varepsilon z(t;\eta(\cdot)) + \mathcal{O}(t,\varepsilon) \quad \forall\, t \text{ in } [t_0,t_f]. \tag{4.85}
$$

Using this and (4.52), one can expand (4.84), as in Sections 3.3 and 3.4, to obtain

$$
\begin{aligned}
\Delta \hat{J} &= \varepsilon \sum_{i=1}^{N} \delta_i \left[H(x^o(t_i+\varepsilon\delta_i), u^o(t_i+\varepsilon\delta_i)+\eta(t_i+\varepsilon\delta_i), \lambda(t_i+\varepsilon\delta_i), t_i+\varepsilon\delta_i) \right. \\
&\qquad \left. - H(x^o(t_i+\varepsilon\delta_i), u^o(t_i+\varepsilon\delta_i), \lambda(t_i+\varepsilon\delta_i), t_i+\varepsilon\delta_i) \right] \\
&\quad + \varepsilon \int_I H_u(x^o(t), u^o(t), \lambda(t), t)\eta(t)dt + \int_{t_0}^{t_f} \mathcal{O}(t;\varepsilon)dt + \mathcal{O}(\varepsilon), \qquad (4.86)
\end{aligned}
$$

where the interval I is defined in (4.82). All the functions have been assumed continuously differentiable in all their arguments. In this way we can find a weak variation as $u(\cdot)+\varepsilon\eta(\cdot) \in \mathcal{U}_T$ that satisfies the variation in the terminal constraints. It follows that (4.86) is nonnegative to first order in ε if (4.80) holds. Note that (4.80) implies $H_u(x^o(t), u^o(t), \lambda(t), t) = 0$. We have thus proved the second part of Theorem 4.4.1, stated below in this section. However, this assumed continuous differentiability of $H(x^o(t), u^o(t), \lambda(t), t)\eta(t)$ is more restrictive than needed to prove strong first-order optimality with terminal constraints.

We now assume that $H(x,u,\lambda,t)$ is differentiable in x, λ, and t but only continuous in u. That is, $L(x,u,t)$ and $f(x,u,t)$ are assumed continuous and not differentiable in u, but $u(\cdot)+\eta(\cdot) \in \mathcal{U}_T$.

The objective is to remove the assumed differentiability of H with respect to u in Equation (4.86). The change in $\hat{J}$ due to changing $u^o(\cdot)$ to $u^o(\cdot) + \eta(\cdot; \varepsilon) \in \mathcal{U}_T$, for $\eta(\cdot; \varepsilon)$ defined in (4.82), is given in (4.84). Due to the continuous differentiability of f with respect to x, we again obtain the expansion (4.85). Using this expansion of the state, the change $\Delta \hat{J}$ becomes

$$
\begin{aligned}
\Delta \hat{J} &= \varepsilon \sum_{i=1}^{N} \delta_i [H(x^o(t_i + \varepsilon\delta_i), u^o(t_i + \varepsilon\delta_i) + \eta(t_i + \varepsilon\delta_i), \lambda(t_i + \varepsilon\delta_i), t_i + \varepsilon\delta_i) \\
&\qquad - H(x^o(t_i + \varepsilon\delta_i), u^o(t_i + \varepsilon\delta_i), \lambda(t_i + \varepsilon\delta_i), t_i + \varepsilon\delta_i)] \\
&\quad + \int_I [H(x^o(t), u^o(t) + \varepsilon\eta(t), \lambda(t), t) - H(x^o(t), u^o(t), \lambda(t), t) \\
&\qquad + \varepsilon H_x(x^o(t), u^o(t), \lambda(t), t) z(t; \eta(\cdot)) + \varepsilon \dot{\lambda}^T(t) z(t; \eta(\cdot))] dt \\
&\quad + \varepsilon [\phi_x(x^o(t_f)) + \nu^T \psi_x(x^o(t_f)) - \lambda(t_f)^T] z(t_f; \eta(\cdot)) \\
&\quad + \int_{t_0}^{t_f} \mathcal{O}(t; \varepsilon) dt + \mathcal{O}(\varepsilon).
\end{aligned} \tag{4.87}
$$

In the above we used the result

$$
\varepsilon H_x(x^o(t), u^o(t) + \varepsilon\eta(t), \lambda(t), t) z(t; \eta(\cdot)) = \varepsilon H_x(x^o(t), u^o(t), \lambda(t), t) z(t; \eta(\cdot)) + \mathcal{O}(\varepsilon), \tag{4.88}
$$

deduced by continuity. Let

$$
\dot{\lambda}(t) = -H_x^T(x^o(t), u^o(t), \lambda(t), t), \quad \lambda(t_f) = \phi_x(x^o(t_f)) + \nu^T \psi_x(x^o(t_f)); \tag{4.89}
$$

then

$$
\begin{aligned}
\delta \hat{J} &= \varepsilon \sum_{i=1}^{N} \delta_i \left[H(x^o(t_i + \varepsilon\delta_i), u^o(t_i + \varepsilon\delta_i) + \eta(t_i + \varepsilon\delta_i), \lambda(t_i + \varepsilon\delta_i), t_i + \varepsilon\delta_i) \right. \\
&\qquad \left. - H(x^o(t_i + \varepsilon\delta_i), u^o(t_i + \varepsilon\delta_i), \lambda(t_i + \varepsilon\delta_i), t_i + \varepsilon\delta_i) \right] \\
&\quad + \int_I \left[H(x^o(t), u^o(t) + \varepsilon\eta(t), \lambda(t), t) - H(x^o(t), u^o(t), \lambda(t), t) \right] dt.
\end{aligned} \tag{4.90}
$$

For $u^o(\cdot) + \eta(\cdot; \varepsilon) \in \mathcal{U}_T$, a sufficient condition for $\delta \hat{J} \geq 0$ is that

$$H(x^o(t), u(t), \lambda(t), t) - H(x^o(t), u^o(t), \lambda(t), t) \geq 0 \tag{4.91}$$

or

$$H(x^o(t), u^o(t), \lambda(t), t) \leq H(x^o(t), u(t), \lambda(t), t) \tag{4.92}$$

for $u^o(\cdot), u(\cdot) \in \mathcal{U}_T$. This completes the proof of Theorem 4.4.1, which is stated below.

Theorem 4.4.1 *Suppose that Assumptions* 3.2.4, 3.3.2, 3.4.1, 4.2.1, *and* 4.2.4 *are satisfied. Suppose further that $u^o(\cdot) \in \mathcal{U}_T$ minimizes the performance criterion* (3.2) *subject to* (3.41), *and* (4.38). *Then there exists a p-vector ν such that*

$$H\left(x^o(t), u^o(t), \lambda(t), t\right) \leq H\left(x^o(t), u(t), \lambda(t), t\right) \quad \forall\, t \text{ in } [t_0, t_f], \tag{4.93}$$

where

$$-\dot{\lambda}^T(t) = H_x(x^o(t), u^o(t), \lambda(t), t), \tag{4.94}$$

$$\lambda^T(t_f) = \phi_x(x^o(t_f)) + \nu^T \psi_x(x^o(t_f)), \tag{4.95}$$

and

$$H(x, u, \lambda, t) \triangleq L(x, u, t) + \lambda^T f(x, u, t). \tag{4.96}$$

Moreover, the above condition is sufficient for J to be strongly first-order optimal at $u^o(\cdot)$.

Since a sufficiency condition for strong first-order optimality is sought, then only Assumption 4.2.1 is required rather than an explicit assumption on controllability. Rigorous proofs of the necessity of Pontryagin's Principle are available in [44, 10, 25, 38], and upon assuming normality, the above conditions result. The second part of the theorem is proved just before the theorem statement.

4.4.1 Strong First-Order Optimality with Control Constraints

If $u(t)$ is required to satisfy (3.106), this does not introduce any difficulty into our straightforward variational proof that Pontryagin's Principle is sufficient for strong first-order optimality. This follows because if (4.93) holds for $u^o(\cdot)$, $u(t) \in \mathcal{U}_T$, then the terms under the summation sign in (4.90) are nonnegative. The latter follows from (4.93) because

$$\begin{aligned} &H(x^o(t), u^o(t) + \varepsilon\eta(t), \lambda(t), t) - H(x^o(t), u^o(t), \lambda(t), t) \\ &\quad = \varepsilon H_u(x^o(t), u^o(t), \lambda(t), t)\eta(t) + \mathcal{O}(\varepsilon), \end{aligned} \tag{4.97}$$

where $u^o(t)$, $u^o(t) + \varepsilon\eta(t) \in \mathcal{U}_T$, and the first-order term dominates for small ε.

The difficulty, described in Section 4.2.3, of deriving necessary conditions for optimality when terminal constraints are present is increased by the presence of (3.106) and the techniques used in [44, 10, 25, 38] are essential. Again, we define the set of bounded piecewise continuous control that meets the terminal constraints as $\mathcal{U}_{BT} \subset \mathcal{U}_T$. In view of the above, we can state the following theorem.

Theorem 4.4.2 *Suppose Assumptions* 3.2.4, 3.3.2, 3.4.1, 4.2.1, *and* 4.2.4 *are satisfied. Suppose further that $u^o(\cdot)$ minimizes the performance criterion* (3.2) *subject to* (3.41), (4.38) *and* (3.106). *Then there exists a p-vector ν such that*

$$H(x^o(t), u^o(t), \lambda(t), t) \le H(x^o(t), u(t), \lambda(t), t) \quad \forall\, t \in [t_0, t_f], \tag{4.98}$$

where

$$u^o(t), u(t) \in \mathcal{U}_{BT}, \tag{4.99}$$

and $\lambda(\cdot)$ *and* H *are as in* (4.94) *and* (4.96). *Moreover, the above condition is sufficient for* J *to be strongly first-order optimal at* $u^o(\cdot)$.

Remark 4.4.1 *We have considered bounds only on the control variable. The extension to mixed control and state constraints is straightforward and can be found in* [11]. *Problems with state-variable inequality constraints are more complex and are beyond the scope of this book. However, necessary conditions for optimality with state-variable inequality constraints can be found in* [30].

Example 4.4.1 (Rocket launch example, revisited) *We first return to the example of Section* 4.3.1 *to see what further insights Theorem* 4.4.2 *gives. Theorem* 4.4.2 *states that* H *should be minimized with respect to the control, subject to* (4.99). *In this example the control is unrestricted so minimization of* H *implies the necessary condition*

$$H_u(x,u,\lambda,t) = -\lambda_3 T \sin u + \lambda_4 T \cos u = 0, \tag{4.100}$$

which implies that

$$\tan u^o(t) = \frac{\lambda_4(t)}{\lambda_3(t)} = -\lambda_4(t) \tag{4.101}$$

because of (4.61). *Now*

$$\begin{aligned} H_{uu}(x,u,\lambda,t) &= -\lambda_3 T \cos u - \lambda_4 T \sin u \\ &= T\cos u - \lambda_4 T \sin u. \end{aligned} \tag{4.102}$$

Using the fact that, from (4.101),

$$\sin u^o = -\lambda_4 \cos u^o, \tag{4.103}$$

we see that

$$\begin{aligned} H_{uu}(x, u^o, \lambda, t) &= T\cos u^o + T\lambda_4^2 \cos u^o \\ &= T(1+\lambda_4^2)\cos u^o > 0 \ \ \textit{for} \ \ -\frac{\pi}{2} < u^o < \frac{\pi}{2}. \end{aligned} \tag{4.104}$$

Inequality (4.104) *implies that* H *has a local minimum with respect to* u *when* $-\frac{\pi}{2} < u^o < \frac{\pi}{2}$ *which satisfies* (4.100). *This local minimum value of* H*, ignoring the terms in* (4.59) *that do not depend on* u*, is obtained when* u^o *is substituted into*

$$\tilde{H}(u) \triangleq -T\cos u + \lambda_4 T \sin u, \tag{4.105}$$

viz.,

$$\tilde{H}(u^o) = -T(1+\lambda_4^2)\cos u^o < 0. \tag{4.106}$$

Now examine

$$\begin{aligned} \tilde{H}^2 + \tilde{H}_u^2 &= \tilde{H}^2 + H_u^2 \\ &= (-T\cos u + \lambda_4 T \sin u)^2 + (T\sin u + \lambda_4 T\cos u)^2 \\ &= T^2(1+\lambda_4^2), \end{aligned} \tag{4.107}$$

which is independent of u^o. *From* (4.107) *we deduce that* $\tilde{H}^2$ *achieves its global maximum value when* $H_u = 0$*, i.e.,* (4.100) *holds. However, because* (4.106) *corresponds to* $\tilde{H}$ *taking on a negative value,* u^o*, given by* (4.101)*, globally minimizes* $\tilde{H}$ *and hence also* $H(x, u, \lambda, t)$. *The conclusion is that* (4.63) *satisfies* (4.98).

Example 4.4.2 *Our next illustrative example has a weak first-order optimum but not a strong first-order optimum. Consider*

$$J(u(\cdot); x_0) = \min_u \int_{t_0}^{t_f} u^3(t)dt \tag{4.108}$$

subject to

$$\dot{x}(t) = u(t), \quad x(t_0) = 0, \tag{4.109}$$

and

$$x(t_f) = 0, \tag{4.110}$$

where x and u are scalars. Here

$$H(x, u, \lambda, t) = u^3 + \lambda u \tag{4.111}$$

and

$$-\dot{\lambda}(t) = 0, \quad \lambda(t_f) = \nu. \tag{4.112}$$

From (4.111)

$$H_u(x, u, \lambda, t) = 3u^2 + \lambda, \tag{4.113}$$

and taking $\nu = 0$ we see that

$$u^o(t) = 0 \ \forall \ t \ \text{ in } \ [t_0, t_f] \tag{4.114}$$

causes

$$H_u(x^o(t), u^o(t), \lambda(t), t) = 0 \ \forall \ t \ \text{ in } \ [t_0, t_f]. \tag{4.115}$$

It follows from Theorem 4.3.1 *that J is weakly first-order optimal at $u^o(\cdot)$.*

Note, however, that $u(t)$ does not minimize $H(x(t), u(t), \lambda(t), t)$, which has no minimum with respect to $u(t)$ because of the cubic term in $u(t)$. Theorem 4.4.2 *then implies that J is not strongly first-order optimal at $u^o(\cdot)$. Stated explicitly,* $\min_u H$

as $u \to -\infty \Rightarrow J \to -\infty$. *This is easily confirmed directly by using the strong perturbation*

$$\eta(t;\varepsilon) = \begin{cases} -(t_f - t_0 - \varepsilon) & \text{for } t_0 \le t \le t_0 + \varepsilon, \\ \varepsilon & \text{for } t_0 + \varepsilon < t \le t_f \end{cases} \tag{4.116}$$

that is added to $u^o(t) = 0$. *This perturbation clearly causes* $\xi(t_f;\varepsilon)$ *to be zero for all* $0 < \varepsilon < t_f - t_0$ *so that* (4.110) *is maintained. Furthermore,*

$$\begin{aligned} J(u^o(\cdot) + \eta(\cdot;\varepsilon); x_0) &= \int_{t_0}^{t_f} \eta^3(t;\varepsilon)dt \\ &= -\int_{t_0}^{t_0+\varepsilon} (t_f - t_0 - \varepsilon)^3 dt + \int_{t_0+\varepsilon}^{t_f} \varepsilon^3 dt \\ &= -(t_f - t_0 - \varepsilon)^3\varepsilon + (t_f - t_0 - \varepsilon)\varepsilon^3 \\ &= (t_f - t_0 - \varepsilon)\varepsilon[\varepsilon^2 - (t_f - t_0 - \varepsilon)^2], \end{aligned} \tag{4.117}$$

which, for small $0 < \varepsilon \ll 1$, $J = -\varepsilon(t_f - t_0)^3 + \mathcal{O}(\varepsilon)$, *verifying that* (4.108) *is not strongly first-order optimal at* $u^o(\cdot)$. *As* (4.110) *is satisfied for all values of* ε, $0 < \varepsilon < t_f - t_0$, *we can also conclude that* (4.108) *does not have a strong local minimum at* $u^o(\cdot)$.

4.5 Unspecified Final Time t_f

As in Section 3.6, we here allow ϕ and ψ to depend explicitly on the unspecified final time t_f, viz., we write these functions as $\phi(x(t_f), t_f)$ and $\psi(x(t_f), t_f)$ and assume that the first partial derivatives are continuous, i.e., as in the next assumption.

Assumption 4.5.1 *The function* $\phi(\cdot,\cdot)$ *and the* p*-dimensional function* $\psi(\cdot,\cdot)$ *are once continuously differentiable in* x *and* t.

We now derive a sufficient condition for

$$J(u(\cdot); x_0, t_f) = \phi(x(t_f), t_f) + \int_{t_0}^{t_f} L(x(t), u(t), t)dt \tag{4.118}$$

to be strongly first-order optimal at $u^o(\cdot)$, t_f^o, subject to the constraint that (4.38) is satisfied. Note that weak and strong optimality are equivalent in the minimization of J with respect to the parameter t_f.

We perturb $u^o(\cdot)$ to $u^o(\cdot)+\eta(\cdot;\varepsilon)$ with $\eta(\cdot;\varepsilon)$ defined in (4.82) and t_f^o to $t_f^o+\varepsilon\Delta$. If $\Delta > 0$, we define $u^o(t)+\eta(t;\varepsilon)$ in the interval $(t_f^o, t_f^o+\varepsilon\Delta]$ as any continuous function with values in $\mathcal{U}_T$, and we require that the perturbed control causes

$$\psi(x(t_f), t_f) = 0 \tag{4.119}$$

at $t_f = t_f^o + \varepsilon\Delta$.

Upon defining

$$\begin{aligned}\hat{J}(u(\cdot);\lambda(\cdot),\nu,x_0,t_f) &= \int_{t_0}^{t_f}\left[H(x(t),u(t),\lambda(t),t)+\dot{\lambda}^T(t)x(t)\right]dt \\ &\quad + \phi(x(t_f),t_f)+\nu^T\psi(x(t_f),t_f)+\lambda^T(t_0)x_0-\lambda^T(t_f)x(t_f),\end{aligned} \tag{4.120}$$

we obtain, using (4.94), that (remembering η from (4.82))

$$\begin{aligned}&\hat{J}(u^o(\cdot)+\eta(\cdot;\varepsilon);\lambda(\cdot),\nu,x_0,t_f^o+\varepsilon\Delta)-\hat{J}(u^o(\cdot);\lambda(\cdot),\nu,x_0,t_f^o)\\ &\quad = \varepsilon\sum_{i=1}^{N}\delta_i\,[H(x^o(t_i+\varepsilon\delta_i),u^o(t_i+\varepsilon\delta_i)+\eta(t_i+\varepsilon\delta_i),\lambda(t_i+\varepsilon\delta_i),t_i+\varepsilon\delta_i)\\ &\qquad\qquad - H(x^o(t_i+\varepsilon\delta_i),u^o(t_i+\varepsilon\delta_i),\lambda(t_i+\varepsilon\delta_i),t_i+\varepsilon\delta_i)]\\ &\quad\quad +\varepsilon\int_I H_u(x^o(t),u^o(t),\lambda(t),t)\eta(t)dt\\ &\quad\quad +\varepsilon\Delta\left[\phi_{t_f}(x^o(t_{f+}^o),t_{f+}^o)+\nu^T\psi_{t_f}(x^o(t_{f+}^o),t_{f+}^o)\right.\\ &\qquad\qquad \left.+\ H(x^o(t_{f+}^o),u^o(t_{f+}^o)+\eta(t_{f+}^o;\varepsilon),\lambda(t_{f+}^o),t_{f+}^o)\right]\\ &\quad\quad +\int_{t_0}^{t_f}\mathcal{O}(t;\varepsilon)dt+\mathcal{O}(\varepsilon),\end{aligned} \tag{4.121}$$

where t_{f+}^o denotes the constant immediately to the right of t_f^o (i.e., the limit is taken of the functions as $t \to t_f^o$ from above). We can now state and prove the following theorem.

Theorem 4.5.1 *Suppose that Assumptions* 3.3.2, 3.4.1, 4.2.1, *and* 4.5.1 *are satisfied. Suppose further that* $u^o(\cdot)$, t_f^o *minimize the performance criterion* (4.118) *subject to* (3.41), (3.106) *and* (4.119). *Then there exists a* p*-vector* ν *such that*

$$H(x^o(t), u^o(t), \lambda(t), t) \le H(x^o(t), u(t), \lambda(t), t) \quad \forall t \in [t_0, t_f] \tag{4.122}$$

and the transversality condition

$$\begin{aligned} \Omega(x^o(t_f^o), u^o(t_f^o), \nu, t_f^o) &\triangleq \phi_{t_f}(x^o(t_f^o), t_f^o) + \nu^T \psi_{t_f}(x^o(t_f^o), t_f^o) \\ &\quad + H(x^o(t_f^o), u^o(t_f^o), \lambda(t_f^o), t_f^o) = 0, \end{aligned} \tag{4.123}$$

where

$$u^o(t), u(t) \in \mathcal{U}_{BT} \tag{4.124}$$

and

$$H(x(t), u(t), \lambda(t), t) = L(x(t), u(t), t) + \lambda^T(t) f(x(t), u(t), t), \tag{4.125}$$

$$\begin{aligned} -\dot{\lambda}^T(t) &= H_x(x^o(t), u^o(t), \lambda(t), t), \\ \lambda^T(t_f^o) &= \phi_x(x^o(t_f^o), t_f^o) + \nu^T \psi_x(x^o(t_f^o), t_f^o). \end{aligned} \tag{4.126}$$

Moreover, the above conditions are sufficient for J *to be strongly first-order optimal at* $u^o(\cdot)$, t_f^o.

Proof: For $\Delta > 0$, (4.122) implies that the summation and the integral term in (4.121) are nonnegative and that, using (4.122), (4.123), and the continuity of $H(x^o(\cdot), u^o(\cdot), \lambda(\cdot), \cdot)$, the term in $\varepsilon\Delta$ is also nonnegative. For $\Delta < 0$ (4.121) is slightly different insofar as the $\varepsilon\Delta$ term is concerned, viz., it becomes

$$-\varepsilon\Delta \left[\phi_{t_f}(x^o(t_{f-}^o), t_{f-}^o) + \nu^T \psi_{t_f}(x^o(t_{f-}^o), t_{f-}^o) + H(x^o(t_{f-}^o), u^o(t_{f-}^o), \lambda(t_{f-}^o), t_{f-}^o)\right].$$

Still, though, the conclusion is unchanged. ▮

Rigorous proofs of the necessity of Pontryagin's Principle are available in [44, 10, 25, 38]. Upon assuming normality, the above conditions result.

4.6 Minimum Time Problem Subject to Linear Dynamics

Of special interest is the linear minimum time problem: Minimize the time t_f (i.e., $\phi(x(t_f), t_f) = t_f$) to reach the terminal point

$$\psi(x(t_f), t_f) = x(t_f) = 0 \tag{4.127}$$

subject to the linear dynamic system

$$\dot{x}(t) = A(t)x(t) + B(t)u(t), \quad x(t_0) = x_0, \tag{4.128}$$

and the control constraint

$$-1 \le u_i(t) \le 1 \;\forall\, t \text{ in } [t_0, t_f], \quad i = 1, \dots, m. \tag{4.129}$$

In this special case, the variational Hamiltonian is

$$H(x, u, \lambda, t) = \lambda^T(t)A(t)x(t) + \lambda^T(t)B(t)u(t). \tag{4.130}$$

From (4.130) we obtain

$$-\dot{\lambda}^T(t) = \lambda^T(t)A(t), \quad \lambda^T(t_f^o) = \nu^T, \tag{4.131}$$

and

$$u_i^o(t) = -\text{ sign } \left[B_i^T(t)\lambda(t)\right] \forall\, t \in [t_0, t_f^o], \quad i = 1, \dots, m, \tag{4.132}$$

where sign $[\sigma]$ is defined in (3.116). The optimal control $u^o(\cdot)$ which satisfies (4.132) is referred to as a "bang-bang" control. It follows that (4.132) is well defined if (the controllability) Assumption 4.2.2 holds. We also have the important condition that, from (4.123),

$$1 + \nu^T B(t_f^o) u^o(t_f^o) = 0. \tag{4.133}$$

Pontryagin's Principle thus states that if a pair $u^o(\cdot)$, t_f^o minimizes

$$J(u(\cdot); x_0) \stackrel{\triangle}{=} t_f \tag{4.134}$$

subject to (4.127)–(4.129), then there exists a ν such that (4.131), (4.132), and (4.133) are satisfied. In general, ν has to be determined numerically.

Example 4.6.1 (Minimum time to the origin: The Bushaw problem) *By way of illustration we particularize* (4.128) *and* (4.129) *to the Bushaw problem* [11]

$$\begin{aligned} \dot{x}_1(t) &= x_2(t), \quad x_1(0) = x_{10}, \quad x_1(t_f) = 0, \\ \dot{x}_2(t) &= u(t), \quad x_2(0) = x_{20}, \quad x_2(t_f) = 0, \end{aligned} \tag{4.135}$$

and

$$-1 \le u(t) \le 1. \tag{4.136}$$

In this case

$$\begin{aligned} -\dot{\lambda}_1(t) &= 0, \qquad \lambda_1(t_f^o) = \nu_1, \\ -\dot{\lambda}_2(t) &= \lambda_1(t), \quad \lambda_2(t_f^o) = \nu_2, \end{aligned} \tag{4.137}$$

so that

$$\begin{aligned} u^o(t) &= -\operatorname{sign} \lambda_2(t) \\ &= -\operatorname{sign} [\nu_2 + (t_f^o - t)\nu_1], \end{aligned} \tag{4.138}$$

and

$$1 + u^o(t_f^o)\nu_2 = 0. \tag{4.139}$$

Conditions (4.138) *and* (4.139) *imply that* ν_2 *is* ± 1.

One can also immediately see from (4.138) *that if* $u^o(\cdot)$ *switches during the interval* $[t_0, t_f^o]$ *it can switch only once. Thus,* $u^o(\cdot)$ *either is constant* (± 1) *for all* t *in* $[t_0, t_f^o]$ *or is* ± 1 *in the interval* $[t_0, t_s]$ *and* ∓ 1 *in the interval* $(t_s, t_f]$*, where* t_s *is the switch time. If the origin cannot be reached from the given initial condition* x_0 *using a constant control* (± 1)*, then this option is ruled out and the one-switch bang-bang control is the only possibility. Upon examining the solutions of* (4.135) *with this form of switching control, it quickly becomes evident whether* $u^o(\cdot)$ *should be* $+1$ *or* -1 *on* $[t_0, t_s]$*. The switch time,* t_s*, can then be calculated easily to ensure that* $x(t_f^o) = 0$*. Then* (4.139) *can be verified. This is best demonstrated in the phase portrait shown in Figure* 4.2*, where the switch curves are parabolas* ($x_1 = -x_2^2/2$ for $u = -1$ *and* $x_1 = x_2^2/2$ for $u = 1$)*. The other trajectories are translated parabolas.*

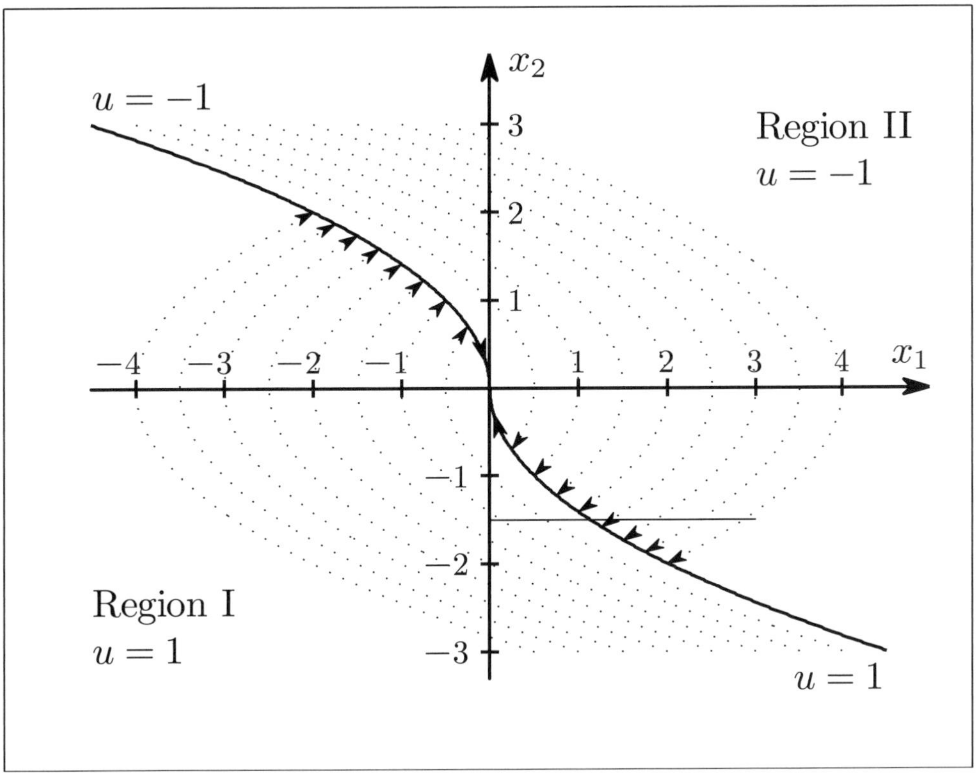

Figure 4.2: Phase portrait for the Bushaw problem.

4.7 Sufficient Conditions for Global Optimality: The Hamilton–Jacobi–Bellman Equation

When terminal equality constraints (4.119) are present and t_f can be specified or free, the Hamilton–Jacobi–Bellman (H-J-B) equation (3.123) can be generalized to

$$-V_t(x(t),t) = \min_{u(t)\in\mathcal{U}_{BT}} \left[L(x(t),u(t),t) + V_x(x(t),t)f(x(t),u(t),t)\right], \qquad (4.140)$$

$$V(x(t_f),t_f) = \phi(x(t_f),t_f) \qquad (4.141)$$

for all $x(t_f)$ and t_f such that $\psi(x(t_f),t_f) = 0$ and is unspecified for all $x(t_f)$ and t_f such that $\psi(x(t_f),t_f) \neq 0$. We then have the following generalization of Theorem 3.5.1.

Theorem 4.7.1 *Suppose there exists a once continuously differentiable function* $V(\cdot,\cdot)$ *of* x *and* t, $t \in [t_0, t_f]$, *that satisfies* (4.140) *and is equal to* $\phi(x(t_f),t_f)$ *when* $\psi(x(t_f),t_f) = 0$. *Suppose further that the control* $u^o(x(t),t)$ *that minimizes*

$$\mathcal{H}(x,u,V_x,t) \stackrel{\triangle}{=} L(x(t),u(t),t) + V_x(x(t),t)f(x(t),u(t),t) \qquad (4.142)$$

subject to the constraint $u(t) \in \mathcal{U}_{BT}$ *is such that Assumption* 3.3.2 *is satisfied. Then, under Assumptions* 3.4.1, 4.2.1, *and* 4.5.1 *the control function* $u^o(x^o(\cdot),\cdot)$ *minimizes* (4.118) *subject to* (3.41), (3.106), *and* (4.119) *and* $V(x_0,t_0)$ *is equal to the minimum value of* (4.118).

Proof: The proof follows as in Section 3.5, in particular that of Theorem 3.5.1.

If we now draw $u(\cdot)$, $u(\cdot) \in \mathcal{U}_{BT}$, assuming (4.140) and (4.141) hold, then in the proof of Theorem 3.5.1, the value of the cost criterion $\hat{J}(u(\cdot);x_0)$ is still expressed by (3.140). Due to (4.140), the integrand of (4.141) is nonnegative and takes on its minimum value of zero when $u(t) = u^o(x(t),t) \in \mathcal{U}_{BT}$, thus completing the proof. ∎

We illustrate the above theorem using the formulation in Section 4.2.1 with $D = I$. In this case the H-J-B equation is

$$-V_t(x(t),t) = \min_{u(t)} \left\{ u^T(t)u(t) + V_x(x(t),t)\left[A(t)x(t) + B(t)u(t)\right] \right\}. \qquad (4.143)$$

If we set

$$\begin{aligned} V(x(t),t) &= 2x^T(t)\Phi^T(t_f,t)W^{-1}(t_0,t_f)\Phi(t_f,t_0)x_0 \qquad (4.144)\\ &- \int_t^{t_f} x_0^T\Phi^T(t_f,t_0)W^{-1}(t_0,t_f)\Phi(t_f,\tau)B(\tau)B^T(\tau)\Phi^T(t_f,\tau)W^{-1}(t_0,t_f)\Phi(t_f,t_0)x_0 d\tau, \end{aligned}$$

we see that at $t = t_f$, $V(x(t_f),t_f) = 0$ when $x(t_f) = 0$. As $\phi(x(t_f)) = 0$, this is as required by Theorem 4.7.1.

Minimizing the right-hand side of (4.143) with respect to $u(t)$ yields

$$u^o(t) = -B^T(t)\Phi^T(t_f,t)W^{-1}(t_0,t_f)\Phi(t_f,t_0)x_0, \qquad (4.145)$$

which by (4.7) drives the linear dynamic system to the origin at $t = t_f$, also as required by Theorem 4.7.1. The right-hand side of (4.143) then becomes

$$\begin{aligned} &2x^T(t)A^T(t)\Phi^T(t_f,t)W^{-1}(t_0,t_f)\Phi(t_f,t_0)x_0 \qquad (4.146)\\ &- x_0^T\Phi^T(t_f,t_0)W^{-1}(t_0,t_f)\Phi(t_f,t)B(t)B^T(t)\Phi^T(t_f,t)W^{-1}(t_0,t_f)\Phi(t_f,t_0)x_0, \end{aligned}$$

which by (4.144) is just $-V_t(x(t),t)$; i.e., the H-J-B equation is satisfied. This proves by Theorem 4.7.1 that (4.145) is optimal. The minimum value of

$$J(u(\cdot);x_0) = \int_{t_0}^{t_f} u^T(t)u(t)dt \qquad (4.147)$$

is then given by $V(x_0,t_0)$ which, from (4.144) and (4.4), reduces to

$$V(x_0,t_0) = x_0^T\Phi^T(t_f,t_0)W^{-1}(t_0,t_f)\Phi(t_f,t_0)x_0. \qquad (4.148)$$

Note, for completeness, that $V_x(x(t),t)$ is just the $\lambda^T(t)$ given by (4.36).

Naturally, in more complicated problems it may be difficult to find an appropriate solution of (4.140) and (4.141). However, this is the price that one pays for attempting to derive a globally optimal control function for nonlinear problems with terminal constraints.

Example 4.7.1 (Minimum time to the origin: Bushaw problem continued) *Some additional insights are obtained by calculating the cost to go from a given point to the origin using the optimal control, i.e., the optimal value function* $V(x_1, x_2, t)$, *for the Bushaw problem presented in Section* 4.6.1. *The time,* t_1, *to reach the switch curve from a point* (x_{01}, x_{02}) *off the switch curve in Region* I *(see Figure* 4.2*) is determined from*

$$\frac{t_1^2}{2} + x_{01}t_1 + x_{02} = -\frac{(t_1 + x_{02})^2}{2}. \tag{4.149}$$

Once t_1 *is obtained, we can obtain the* $x_{s1} = \frac{t_1^2}{2} + x_{01}t_1 + x_{02}$ *and* $x_{s2} = t_1 + x_{02}$ *on the switching curve, and then the time* t_2 *to the origin given by* $\frac{t_2^2}{2} + x_{s1}t_2 + x_{s2} = 0$. *Summing the two times,* $t_1 + t_2$, *we obtain the optimal value function:*

$$\text{in Region I}: \quad V(x_1, x_2, t) = t - x_2 + [2(x_2^2 - 2x_1)]^{\frac{1}{2}}, \tag{4.150}$$

$$\text{in Region II}: \quad V(x_1, x_2, t) = t + x_2 + [2(x_2^2 + 2x_1)]^{\frac{1}{2}}, \tag{4.151}$$

where x_1, x_2, *and* t *denote an arbitrary starting point. Figure* 4.3 *gives a plot of the optimal value function where* x_2 *is held fixed at* -1.5 *as shown in Figure* 4.2 *and* x_1 *is allowed to vary. Note that in going from Region* I *to* II *at* $x_1 = 1.125$ *the optimal value function is continuous, but* not *differentiable. Therefore, along the switch curve the H-J-B equation is not applicable. However, off the switch curves the H-J-B equation does apply. Note that for small changes in* x_1 *and* x_2 *across the switch curve produces a large change in* ν_2. *Since the transversality condition* $1 + u^o\nu_2 = 0$ *implies that in Region* I *with* $u^o = +1$, *then* $\nu_2 = -1$, *and in Region* II *with* $u^o = -1$, *then* $\nu_2 = 1$.

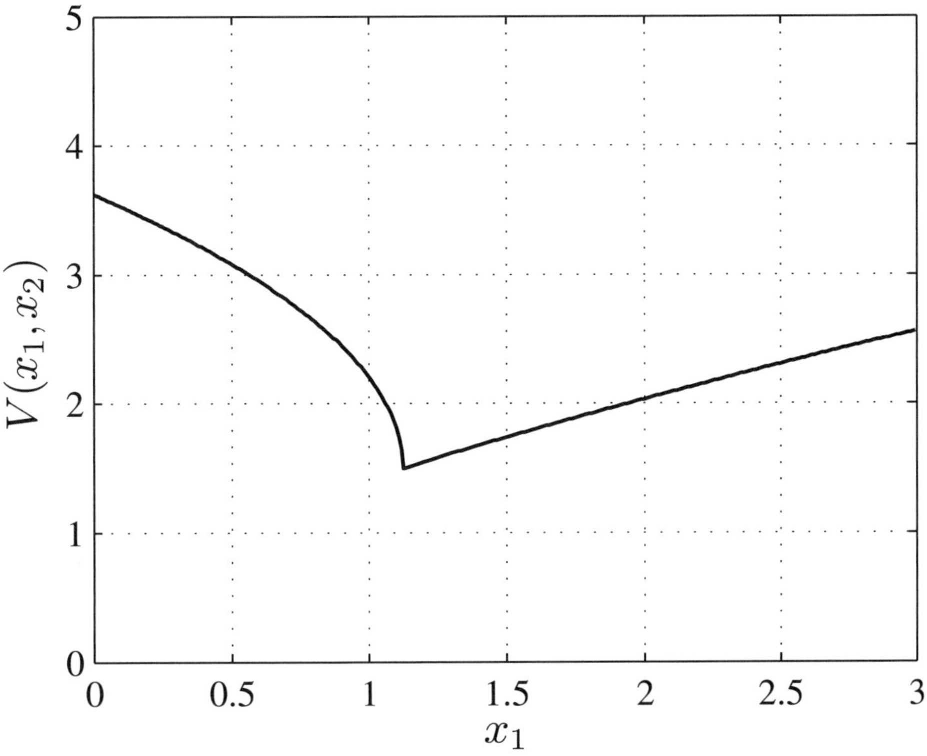

Figure 4.3: Optimal value function for the Bushaw problem.

The optimal value function in the Bushaw example is not differentiable at the switch curves but does satisfy the H-J-B equation everywhere else. In the next chapter a normality condition requires that the system be controllable about any extremal path. Since there are no switches once on the switch curve, the path is not controllable. It has been suggested in [32] that along the switch curve the Lagrange multiplier $\lambda(t)$ is related to the derivative of the unconstrained optimal value function along the switch curve and related to the derivative of the constrained optimal value function in Regions I and II given in (4.150) and (4.151). However, this lack of differentability over a manifold in certain classes of optimization problems has been one drawback for the application of H-J-B theory and the elevation of Pontryagin's Principle [38], which although a local theory does not suffer from this pathological defect.

Problems

1. Consider the problem of finding $u(\cdot)$ that minimizes

$$J = \int_0^3 |u| dt$$

subject to

$$\begin{aligned} &\dot{x}_1 = x_2, \quad x_1(0) = 0, \quad x_1(3) = 2, \\ &\dot{x}_2 = u, \quad x_2(0) = 0, \quad x_2(3) = 0, \\ &|u| \leq 1. \end{aligned}$$

Find the minimizing path.

2. Find the minimum time path, control, and terminal time for

$$\begin{aligned} &\dot{x}_1 = u, \quad x_1(0) = 1, \quad x_1(t_f) = 0, \\ &\dot{x}_2 = \tfrac{u^2}{2}, \quad x_2(0) = 0, \quad x_2(t_f) = 1. \end{aligned}$$

3. Consider a two-stage rocket where, after burnout ($t = t_b$) of the first stage, the second stage is to go into a prescribed orbit and the first stage must return to a prescribed landing site. Let the dynamic equations during the first-stage boost be

$$\dot{x} = f(x, u, t), \quad x(0) = x_0 = \text{ given}, \quad t \in [0, t_b].$$

The dynamics after first-stage burnout ($t = t_b$) are

$$\begin{aligned} &\dot{x}_1 = f_1(x_1, u_1, t), \quad \psi(x_1(t_{f1}), t_{f1}) = 0, \quad t \in (t_b, t_{f1}], \\ &\dot{x}_2 = f_2(x_2, u_2, t), \quad \psi(x_2(t_{f2}), t_{f2}) = 0, \quad t \in (t_b, t_{f2}]. \end{aligned}$$

The problem is to maximize

$$J = \phi(x_2(t_{f2}))$$

subject to the above constraints. At burnout, the states are continuous, i.e.,

$$x(t_b^-) = x_1(t_b^+) = x_2(t_b^+).$$

Determine the first-order necessary conditions for a weak local minimum.

4. Consider the Bushaw problem (free final time) with penalty functions on the terminal states [11]:

$$\min_{t_f,u} \gamma_1 x_{1_f}^2 + \gamma_2 x_{2_f}^2 + t_f, \tag{4.152}$$

where x_{1_0} and x_{2_0} are given. The state equations are

$$\dot{x}_1 = x_2, \tag{4.153}$$

$$\dot{x}_2 = u, \tag{4.154}$$

$$|u| \leq 1. \tag{4.155}$$

Determine the optimal final time and the optimal control history. Show that the solution converges to the solution to the Bushaw problem as $\gamma_1 \to \infty$ and $\gamma_2 \to \infty$.

CHAPTER 5

Second-Order Local Optimality: The Linear Quadratic Control Problem

Introduction

Probably the most used result in optimal control theory is that of the solution to the linear quadratic (LQ) problem, where the dynamic equations and terminal constraints are linear and the performance criterion is a quadratic function of the state and control variables. The solution to this problem produces a control variable as a linear function of the state variables. This solution forms the basis of modern control synthesis techniques because it produces controllers for multi-input/multi-output systems for both time-varying and time-invariant systems [11, 1, 34, 2, 42]. Furthermore, this problem also forms the basis of the accessory minimum problem in the calculus of variations [11, 22, 6, 26, 19, 32, 20].

The presentation here unites previous results (see [1, 8, 11, 20, 34], among others) through a derivation which makes explicit use of the symplectic property of Hamiltonian systems. In this way, our derivation of the solution to the LQ problem is very natural.

The LQ problem is especially important for determining additional conditions for local minimality. In particular, the second-order term in the expansion of the augmented cost criterion can be converted into an LQ problem called the Accessory Problem in the Calculus of Variations. A significant portion of this chapter is devoted to determining the necessary and sufficient conditions for the second variation in the cost, evaluated along a local optimal path, to be positive. Conditions are also given for the second variation to be strongly positive. This means that the second variation dominates over all other terms in the expansion of the cost.

5.1 Second Variation: Motivation for the Analysis of the LQ Problem

Let us return to our original nonlinear dynamical system with general performance index, terminal constraints, and free final time. The equation of motion is

$$\dot{x} = f(x, u, t), \quad x(t_0) = x_0 \text{ given.} \tag{5.1}$$

The problem is

$$\min_{u(t)\in\mathcal{U}_T, t_f} J, \quad J \triangleq \phi(x(t_f), t_f) + \int_{t_0}^{t_f} L(x(t), u(t), t)dt, \tag{5.2}$$

subject to the constraint $\psi(x(t_f), t_f) = 0$.

As in earlier chapters, we include the equations of motion in the performance index through the use of a vector $\lambda(\cdot)$ of Lagrange multiplier functions and we append the final state constraints using a vector ν of Lagrange multipliers. The augmented performance index becomes

$$\hat{J} = \phi(x(t_f), t_f) + \nu^T\psi(x(t_f), t_f) + \int_{t_0}^{t_f} \left[H(x(t), u(t), \lambda(t), t) - \lambda^T(t)\dot{x}\right] dt, \tag{5.3}$$

where the Hamiltonian $H(\cdot, \cdot, \cdot, \cdot)$ is defined as usual as $H(x, u, \lambda, t) \triangleq L(x, u, t) + \lambda^T f(x, u, t)$.

Assumption 5.1.1 *The functions $f(\cdot,\cdot,\cdot), L(\cdot,\cdot,\cdot), \phi(\cdot,\cdot)$, and $\psi(\cdot,\cdot)$ are all twice differentiable with respect to their arguments.*

Assumption 5.1.2 *There exists a pair of vector-valued functions and a scalar parameter $(x^o(\cdot), u^o(\cdot), t_f^o)$ that satisfy Equation (5.1) and minimize the performance index subject to the terminal constraints over all admissible functions $u(\cdot)$.*

We assume that we have generated a locally minimizing path by satisfying the first-order necessary conditions, e.g., Theorem 4.5.1, which produce the extremal values $(x^o(\cdot), u^o(\cdot), \lambda^o(\cdot), \nu^o, t_f^o)$. Letting the variations in the state, control, and multipliers be defined as

$$\Delta x \stackrel{\triangle}{=} x - x^o, \quad \Delta\lambda \stackrel{\triangle}{=} \lambda - \lambda^o, \quad \Delta u \stackrel{\triangle}{=} u - u^o, \quad \Delta\nu \stackrel{\triangle}{=} \nu - \nu^o, \quad \Delta t_f \stackrel{\triangle}{=} t_f - t_f^o, \tag{5.4}$$

where Δ means a total variation and requiring that $u(\cdot)$ be an element of the set of admissible control functions, we expand the augmented cost criterion in a Taylor series as

$$\begin{aligned}
\Delta\hat{J} &= \hat{J}(u(\cdot), t_f; x_0, \lambda(\cdot), \nu) - \hat{J}(u^o(\cdot), t_f^o; x_0, \lambda^o(\cdot), \nu^o) \\
&= \int_{t_0}^{t_f^o} \left[\Delta H - \Delta(\lambda^T \dot{x})\right] dt + \Delta\left[\phi + \nu^T\psi\right]\Big|_{t_f = t_f^o} \\
&= \int_{t_0}^{t_f^o} \left[H_x \Delta x + H_u \Delta u - \lambda^T \Delta\dot{x} + (H_\lambda - \dot{x})^T \Delta\lambda\right] dt + \left[\phi_x + \nu^T\psi_x\right]\Big|_{t_f = t_f^o} \Delta x \\
&\quad + \Delta\nu^T \psi\big|_{t_f = t_f^o} + \Omega\big|_{t_f = t_f^o} \Delta t_f + \frac{1}{2}\int_{t_0}^{t_f^o} \left[\Delta x^T H_{xx} \Delta x + 2\Delta x^T H_{xu} \Delta u \right. \\
&\qquad \left. + \Delta u^T H_{uu} \Delta u + 2\left(\Delta x^T H_{x\lambda} \Delta\lambda + \Delta\lambda^T H_{\lambda u} \Delta u - \Delta\lambda^T \Delta\dot{x}\right)\right] dt \\
&\quad + \frac{1}{2}\begin{bmatrix} \Delta x^T & \Delta\nu^T & \Delta t_f \end{bmatrix} \begin{bmatrix} \phi_{xx} + (\nu^T\psi)_{xx} & \psi_x^T & \Omega_x^T \\ \psi_x & 0 & \Omega_\nu^T \\ \Omega_x & \Omega_\nu & \frac{d\Omega}{dt} \end{bmatrix}_{t_f = t_f^o} \begin{bmatrix} \Delta x \\ \Delta\nu \\ \Delta t_f \end{bmatrix} \\
&\quad + \text{H.O.T.},
\end{aligned} \tag{5.5}$$

where Ω is defined in (4.123) and H.O.T. denotes higher-order terms, that is, terms that include three or more variational values multiplied together.

Remark 5.1.1 *There is no term $\Delta\lambda^T H_{\lambda\lambda}\Delta\lambda$ because H is linear in λ by definition.*

Recall the first-order necessary conditions for optimality (essentially from Theorem 4.5.1):

$$\begin{aligned} \dot{x} &= H_\lambda, \\ H_u &= 0, \\ \psi(x^o(t_f^o), t_f^o) &= 0, \\ \dot{\lambda}^T &= -H_x, \quad \lambda^T(t_f^o) = \left[\phi_x + \nu^T \psi_x\right]_{t=t_f^o}, \\ \Omega(x^o(t_f^o), u^o(t_f^o), \nu, t_f^o) &= 0. \end{aligned} \tag{5.6}$$

These cause the first-order terms in the expansion to be zero. As we have shown for $\Delta x(\cdot)$, the variations in $\Delta\lambda(\cdot)$, $\Delta\nu$, and Δt_f can also be expanded as

$$\begin{aligned} \Delta x(\cdot) &= \epsilon z(\cdot) + \mathcal{O}(\cdot, \epsilon), \quad \Delta\lambda(\cdot) = \epsilon\tilde{\lambda}(\cdot) + \mathcal{O}(\cdot, \epsilon), \\ \Delta\nu &= \epsilon\tilde{\nu} + \mathcal{O}(\epsilon), \quad \Delta t_f = \epsilon\Delta, \quad \Delta u(\cdot) = \epsilon\eta(\cdot). \end{aligned} \tag{5.7}$$

Using these expansions and integrating by parts where necessary, the variation in the performance index becomes

$$\begin{aligned} \Delta\hat{J} &= \epsilon\lambda^T(t_0)z(t_0) \\ &+ \epsilon^2 \frac{1}{2}\int_{t_0}^{t_f^o} \left\{ \begin{bmatrix} z^T & \eta^T \end{bmatrix} \begin{bmatrix} H_{xx} & H_{xu} \\ H_{ux} & H_{uu} \end{bmatrix} \begin{bmatrix} z \\ \eta \end{bmatrix} + 2\tilde{\lambda}^T(f_x z + f_u \eta - \dot{z}) \right\} dt \\ &+ \epsilon^2 \frac{1}{2} \begin{bmatrix} z^T & \tilde{\nu}^T & \Delta \end{bmatrix} \begin{bmatrix} \phi_{xx} + (\nu^T\psi)_{xx} & \psi_x^T & \Omega_x^T \\ \psi_x & 0 & \Omega_\nu^T \\ \Omega_x & \Omega_\nu & \frac{d\Omega}{dt} \end{bmatrix}_{t_f = t_f^o} \begin{bmatrix} z \\ \tilde{\nu} \\ \Delta \end{bmatrix} \\ &+ \int_{t_0}^{t_f^o} \mathcal{O}(t, \epsilon^2)dt + \mathcal{O}(\epsilon^2). \end{aligned} \tag{5.8}$$

Since in our original problem the initial state is given, it is clear that $\Delta x(t_0) = z(t_0) = 0$, making the term involving it disappear.

Letting $\epsilon \to 0$, we know that

$$\lim_{\epsilon \to 0} \frac{\mathcal{O}(\cdot, \epsilon^2)}{\epsilon^2} = 0 \tag{5.9}$$

so that

$$\begin{aligned} \delta^2 \hat{J} &\triangleq \lim_{\epsilon \to 0} \frac{\Delta \hat{J}}{\epsilon^2} \\ &= \frac{1}{2} \int_{t_0}^{t_f^o} \left\{ \begin{bmatrix} z^T & \eta^T \end{bmatrix} \begin{bmatrix} H_{xx} & H_{xu} \\ H_{ux} & H_{uu} \end{bmatrix} \begin{bmatrix} z \\ \eta \end{bmatrix} + 2\tilde{\lambda}^T (f_x z + f_u \eta - \dot{z}) \right\} dt \\ &\quad + \frac{1}{2} \begin{bmatrix} z^T & \tilde{\nu}^T & \Delta \end{bmatrix} \begin{bmatrix} \phi_{xx} + (\nu^T \psi)_{xx} & \psi_x^T & \Omega_x^T \\ \psi_x & 0 & \Omega_\nu^T \\ \Omega_x & \Omega_\nu & \frac{d\Omega}{dt} \end{bmatrix}_{t=t_f^o} \begin{bmatrix} z \\ \tilde{\nu} \\ \Delta \end{bmatrix}. \end{aligned} \tag{5.10}$$

This second variation of the cost criterion is required to be positive definite for all variations about the assumed locally minimizing path $(x^o(\cdot), u^o(\cdot), t_f^o)$. If the second variation can be made negative, then there is another path neighboring to the optimal path that will give a smaller value of the cost criterion. However, this would contradict the assumed optimality of $(x^o(\cdot), u^o(\cdot), t_f^o)$ which satisfies only the first-order necessary conditions and would be an extremal path. From the second variation, which is equivalent to an LQ problem, conditions for minimality can be determined. We first find conditions for the fixed-time second variation without terminal constraints to be positive definite in Section 5.4. Then, conditions for positivity of the fixed-time second variation are given in Section 5.5 with terminal constraints and, finally, in Section 5.8 for free-time and with terminal constraints.

As commonly presented, the LQ problem is a fixed-time problem. Therefore, the terminally constrained LQ problem is now presented where $\Delta = 0$. We will return to the more general problem in Section 5.8. Recalling our definitions of Equation (5.7), we expand the equation of motion (5.1) as

$$\dot{x}(\cdot) = \dot{x}^o(\cdot) + \epsilon\dot{z}(\cdot) + \mathcal{O}(\cdot, \epsilon) = f(x, u, t). \tag{5.11}$$

Expanding f about the nominal trajectory gives

$$\begin{aligned}\dot{x}^o(\cdot) + \epsilon\dot{z}(\cdot) + \mathcal{O}(\cdot, \epsilon) &= f(x^o, u^o, \cdot) + f_x(x^o, u^o, \cdot)(\epsilon z + \mathcal{O}(\cdot, \epsilon)) \\ &\quad + f_u(x^o, u^o, \cdot)\epsilon\eta + \text{H.O.T.}.\end{aligned}$$

Subtracting out the zero quantity $\dot{x}^o(\cdot) = f(x^o, u^o, \cdot)$ gives

$$\epsilon\dot{z}(\cdot) = \epsilon f_x(x^o, u^o, \cdot)z(\cdot) + \epsilon f_u(x^o, u^o, \cdot)\eta(\cdot) + \mathcal{O}(\cdot, \epsilon^2), \tag{5.12}$$

where we have noted that all higher-order terms in the expansion include ϵ to the second or higher power. Dividing through by ϵ and again letting $\epsilon \to 0$, we get the equation

$$\dot{z}(\cdot) = f_x(x^o, u^o, \cdot)z(\cdot) + f_u(x^o, u^o, \cdot)\eta(\cdot), \quad z(t_0) = 0. \tag{5.13}$$

The initial condition comes from the requirement that $\Delta x(t_0) = \epsilon z(t_0) + \mathcal{O}(t_0, \epsilon) = 0$.

Consider Equations (5.10) and (5.13). We see that Equation (5.10) is precisely the augmented performance index we would have obtained for the problem

$$\min_{\eta(\cdot)} J(\eta(\cdot)), \tag{5.14}$$

where

$$\begin{aligned}J(\eta(\cdot)) &\triangleq \frac{1}{2}\int_{t_0}^{t_f} \begin{bmatrix} z^T & \eta^T \end{bmatrix} \begin{bmatrix} H_{xx} & H_{xu} \\ H_{ux} & H_{uu} \end{bmatrix} \begin{bmatrix} z \\ \eta \end{bmatrix} dt \\ &\quad + \frac{1}{2}\left[z^T(\phi_{xx} + (\nu^T\psi)_{xx})z\right]_{t=t_f}\end{aligned} \tag{5.15}$$

subject to the equations of motion (5.13) with the terminal constraint

$$\psi_x(x^o(t_f))z(t_f) = 0. \tag{5.16}$$

Further, we note that if the nominal trajectory is truly minimizing, then the solution to this problem must be $\eta(\cdot) \equiv 0$. Otherwise, there would exist some control variation $\epsilon\eta(\cdot)$ and resulting variation $\epsilon z(\cdot)$ in the state history for which the second term in the expansion of the performance index would be less than that for the nominal trajectory. Therefore, the performance index as a whole would be less than that of the nominal trajectory. The minimization of Equation (5.15) with respect to Equations (5.13) and (5.16) is known as the Accessory Minimum Problem in the Calculus of Variations and is formulated as an LQ problem. In the next section the LQ problem is stated with a simpler notation but follows from the second variation developed in this section.

5.2 Preliminaries and LQ Problem Formulation

The problem of minimizing the quadratic performance criterion

$$\begin{aligned} J(u(\cdot); x(t_0), t_0) &= \frac{1}{2}x^T(t_f)S_f x(t_f) + \frac{1}{2}\int_{t_0}^{t_f} [x^T(t)Q(t)x(t) + 2u^T(t)C(t)x(t) \\ &\quad + u^T(t)R(t)u(t)]dt \end{aligned} \tag{5.17}$$

with respect to $u(\cdot)$ and where $t \in [t_0, t_f]$, $x(t) \in \mathbb{R}^n$, and $u(t) \in \mathbb{R}^m$, subject to the linear dynamic constraint

$$\dot{x}(t) = A(t)x(t) + B(t)u(t), \tag{5.18}$$

with initial condition

$$x(t_0) = x_0 \tag{5.19}$$

and terminal constraints

$$Dx(t_f) = 0, \tag{5.20}$$

where D is a $p \times n$ matrix that will be studied in detail. The matrices $Q(t)$, $C(t)$, $R(t)$, $A(t)$, and $B(t)$ are assumed to be piecewise continuous functions of time, and without loss of generality, $Q(t) = Q^T(t)$, $R(t) = R^T(t)$, and $S_f = S_f^T$.

Remark 5.2.1 *Relate the quadratic cost criterion of Equation* (5.15) *with Equation* (5.17)*, the linear dynamics of Equation* (5.13) *with Equation* (5.18)*, and the linear terminal constraint of Equation* (5.16) *with Equation* (5.20)*. Therefore, the solution to the LQ problem of this section is the solution to the second variation or accessory minimum problem, when setting* $x_0 = 0$.

Assumption 5.2.1 *The matrix* $R(t) > 0$ *and bounded for all* t *in the interval* $t_0 \leq t \leq t_f$*. The implication of relaxing this assumption to positive semidefinite* R *is discussed in Section* 5.4.6.

Assumption 5.2.2 *The control function* $u(\cdot)$ *belongs to the class* $\mathcal{U}$ *of piecewise continuous* m*-vector functions of* t *in the interval* $[t_0, t_f]$.

Initially, no additional restrictions are required for $Q(t)$ and S_f other than symmetry. However, in later sections, special but important results are obtained by requiring that $Q(t)$ and S_f be at least positive semidefinite. Furthermore, the initial time, t_0, on occasion throughout this chapter, is considered to be a variable and not a fixed value.

5.3 First-Order Necessary Conditions for Optimality

In this section the first variation of the LQ problem is established. To include the dynamic and terminal constraints explicitly in the cost criterion, $J(u(\cdot); x(t_0), t_0)$ (given by (5.17)) is augmented by adjoining (5.18) by means of a continuously differentiable n-vector function of time $\lambda(\cdot)$ and (5.20) by means of a p-vector, ν, as

$$\begin{aligned}&\hat{J}\left(u(\cdot);\lambda(\cdot),\nu,x_0,t_0\right)\\&\triangleq J\left(u(\cdot);x_0,t_0\right)+\int_{t_0}^{t_f}\lambda^T(t)\left[A(t)x(t)+B(t)u(t)-\dot{x}(t)\right]+\nu^T Dx(t_f).\end{aligned}\tag{5.21}$$

Note that

$$\hat{J}\left(u(\cdot);\lambda(\cdot),\nu,x_0,t_0\right)=J\left(u(\cdot);x_0,t_0\right)\tag{5.22}$$

when (5.18) and (5.20) hold.

For convenience, define the variational Hamiltonian as

$$\begin{aligned}H(x(t),u(t),\lambda(t),t)\ &\triangleq\ \frac{1}{2}\left[x^T(t)Q(t)x(t)+2u^T(t)C(t)x(t)\right.\\&\qquad\left.+\,u^T(t)R(t)u(t)\right]+\lambda^T(t)\left[A(t)x(t)+B(t)u(t)\right].\end{aligned}\tag{5.23}$$

Integration of (5.21) by parts and using (5.23)

$$\begin{aligned}&\hat{J}\left(u(\cdot);\lambda(\cdot),\nu,x_0,t_0\right)\\&\quad=\int_{t_0}^{t_f}\left[H(x(t),\lambda(t)u(t),t)+\dot{\lambda}^T(t)x(t)\right]dt+\ \lambda^T(t_0)x_0-\lambda^T(t_f)x(t_f)\\&\qquad+\frac{1}{2}x^T(t_f)S_f x(t_f)+\nu^T Dx(t_f).\end{aligned}\tag{5.24}$$

Suppose there is a control $u^o(\cdot)\in\mathcal{U}$ that minimizes (5.17) and causes

$$Dx^o(t_f)=0.\tag{5.25}$$

The objective is to determine necessary conditions for which the cost is a minimum. This is done by evaluating changes in $\hat{J}$ brought about by changing $u^o(\cdot)$ to $u(\cdot)\in\mathcal{U}$. Denote the change in the control as $\delta u(t)\triangleq u(t)-u^o(t)$ and the resulting change in the state as $\delta x(t)\triangleq x(t)-x^o(t)$. Following the methodology laid out in section 4.4, we require conditions that guarantee that the change in $\hat{J}$ is nonnegative for all admissible variations. First, the variations are limited to those for which δx remains small, i.e., $\parallel\delta x(t)\parallel\leq\bar{\varepsilon}$ for all $t\in[t_0,t_f]$. However, strong variations in δu are allowed. For example, $\delta u(\cdot)$ can be chosen as

$$\delta u(t) = \varepsilon(t)\eta(t), \tag{5.26}$$

where $\eta(t) \in \mathcal{U}$ and

$$\varepsilon(t) = \begin{cases} 1, & t_i \le t \le t_i + \varepsilon\delta_i, \quad i = 1, \dots, n, \\ \varepsilon, & t \in I \triangleq \left\{ [t_0, t_f] - \bigcup_i [t_i, t_i + \varepsilon\delta_i] \right\}, \end{cases} \tag{5.27}$$

where $\varepsilon > 0$ is sufficiently small and $\{t_i, t_i + \varepsilon\delta_i\} \in [t_0, t_f]$ are arbitrary with small combined length. Therefore,

$$\begin{aligned} &\Delta\hat{J} \triangleq \Delta\hat{J}(u(\cdot), u^o(\cdot); \lambda(\cdot), \nu, x_0, t_0) \\ =\ & \hat{J}\left(u(\cdot); \lambda(\cdot), \nu, x_0, t_0\right) - \hat{J}\left(u^o(\cdot); \lambda(\cdot), \nu, x_0, t_0\right) \\ =\ & \int_{t_0}^{t_f} \Big[x^{o^T}(t)Q(t)\delta x(t) + u^{o^T}(t)C(t)\delta x(t) + x^{o^T}(t)C^T(t)\delta u(t) \\ & \qquad + \frac{1}{2}\delta u^T(t)R(t)\delta u(t) + u^{o^T}(t)R(t)\delta u(t) \\ & \qquad + \dot{\lambda}^T(t)\delta x(t) + \lambda^T(t)A(t)\delta x(t) + \lambda^T(t)B(t)\delta u(t) \Big] dt \\ & - \lambda^T(t_f)\delta x(t_f) + x^{o^T}(t_f)S_f\delta x(t_f) + \nu^T D\delta x(t_f) \\ & + \int_{t_0}^{t_f} \mathcal{O}(t;\varepsilon)dt + \mathcal{O}(\varepsilon), \end{aligned} \tag{5.28}$$

where the function $\mathcal{O}(t;\varepsilon)$ is piecewise continuous in t and

$$\frac{\mathcal{O}(t;\varepsilon)}{\varepsilon} \to 0 \quad \text{as} \quad \varepsilon \to 0 \quad \text{for each} \quad t. \tag{5.29}$$

Now set

$$\begin{aligned} -\dot{\lambda}^T(t) &= x^{o^T}(t)Q(t) + u^{o^T}C(t) + \lambda^T(t)A(t), \\ \lambda^T(t_f) &= x^{o^T}S_f + \nu^T D. \end{aligned} \tag{5.30}$$

For fixed ν this is a legitimate choice for $\lambda(\cdot)$, since (5.30) is a linear differential equation in $\lambda(t)$ with continuous coefficients, having a unique solution [15].

For ε sufficiently small, (5.28), having substituted in (5.30), is rewritten using (5.23) as

$$\Delta\hat{J} = \int_{t_0}^{t_f} \left[\frac{1}{2}\delta u^T(t)R(t)\delta u(t) + H_u(x^o(t), u^o(t), \lambda(t), t)\delta u(t)\right] dt + \int_{t_0}^{t_f} \mathcal{O}(t;\varepsilon)dt + \mathcal{O}(\varepsilon). \tag{5.31}$$

Since for $t \in [t_0, t_f]$ $H_{uu} = R(t) > 0$, the strong form of the classical Legendre–Clebsch condition, the variation in the cost function $\Delta\hat{J}$ can be reduced if the second term is made negative. This is easily done by choosing $\eta(t) \in \mathcal{U}$ as

$$\eta(t) = -R(t)^{-1}H_u(x^o(t), \lambda(t), u^o(t), t)^T. \tag{5.32}$$

Note that in arbitrary small time intervals where $\varepsilon(t) = 1$, this choice minimizes the integral in (5.31).

This choice of η is particularly significant since it can be shown under an additional assumption that there exists a ν such that $\eta(\cdot)$ given by (5.32) causes (5.25) to hold. (Also see Section 4.3.1.)

Assumption 5.3.1 *The $p \times p$ matrix $\bar{W}(t_0, t_f)$ is positive definite where*

$$\bar{W}(t_0, t_f) \triangleq \int_{t_0}^{t_f} D\Phi(t_f, t)B(t)R^{-1}(t)B^T(t)\Phi^T(t_f, t)D^T dt. \tag{5.33}$$

Remark 5.3.1 *This assumption is equivalent to Assumption 4.2.3 and becomes the controllability condition [8] when $p = n$ and D has rank n.*

From the linearity of (5.18), $z(t;\eta(\cdot))$ satisfies the equation

$$\dot{z}(t;\eta(\cdot)) = A(t)z(t;\eta(\cdot)) + B(t)\eta(t), \quad z(t_0;\eta(\cdot)) = 0 \tag{5.34}$$

so that

$$z(t_f;\eta(\cdot)) = \int_{t_0}^{t_f} \Phi(t_f, t)B(t)\eta(t)dt. \tag{5.35}$$

By using (5.23) and (5.32)

$$z(t_f;\eta(\cdot)) = -\int_{t_0}^{t_f} \Phi(t_f,t)B(t)R^{-1}(t)\left[R(t)u^o(t) + C(t)x^o(t) + B^T(t)\lambda(t)\right] dt. \quad (5.36)$$

The linear forced equation (5.30) is solved as

$$\lambda(t) = \Phi^T(t_f,t)S_f x^o(t_f) + \Phi^T(t_f,t)D^T\nu + \int_t^{t_f} \Phi^T(\tau,t)\left(Q(\tau)x^o(\tau) + C^T(\tau)u^o(\tau)\right)d\tau, \quad (5.37)$$

and then, using (5.37) in (5.36), an equation explicit in ν results. Premultiplying (5.36) by D leads to the desired equation that satisfies the terminal constraints in the presence of variations in state and control as

$$\begin{aligned} &Dz(t_f;\eta(\cdot)) \\ &\quad = -\int_{t_0}^{t_f} D\Phi(t_f,t)B(t)R^{-1}(t)\Bigg[R(t)u^o(t) + C(t)x^o(t) + B^T(t)\Phi^T(t_f,t)S_f x^o(t_f) \\ &\qquad + B^T(t)\int_t^{t_f} \Phi^T(\tau,t)\left(Q(\tau)x^o(\tau) + C^T(\tau)u^o(\tau)\right)d\tau\Bigg]dt \\ &\quad - \left[\int_{t_0}^{t_f} D\Phi(t_f,t)B(t)R(t)^{-1}B^T(t)\Phi^T(t_f,t)D^T dt\right]\nu = 0. \end{aligned} \quad (5.38)$$

By Assumption 5.3.1, a unique value of ν can be obtained, independent of ε, which satisfies the constraints (5.20). Consequently, with this choice of $\eta(\cdot)$, (5.22) holds, and the change in J is[7]

$$\Delta J = -\int_{I_n} \frac{1}{2}\|H_u\|^2_{R(t)^{-1}}dt - \varepsilon\int_I \|H_u\|^2_{R(t)^{-1}}dt + \int_{t_0}^{t_f} \mathcal{O}(t;\varepsilon)dt + \mathcal{O}(\varepsilon), \quad (5.39)$$

where $I_n = \bigcup_i [t_i, t_i + \varepsilon\delta_i]$ and the intervals over which the integrals are taken are given in (5.27). Note that $\mathcal{O}(t;\varepsilon)$ includes all higher-order variations such that (5.29) holds. Since the intervals are arbitrary, the control variations are not restricted to

[7] $\|H_u\|^2_{R(t)^{-1}}$ is the norm square of H_u weighted by $R(t)^{-1}$, i.e., $\|H_u\|^2_{R(t)^{-1}} = H_u^T R(t)^{-1} H_u$.

only small variations. Therefore, a necessary condition for the variation in the cost criterion, ΔJ, to be nonnegative for arbitrary strong variations, δu, is that

$$H_u(x^o(t), \lambda(t), u^o(t), \lambda(t), t) = 0. \tag{5.40}$$

The above results are summarized in the following theorem.

Theorem 5.3.1 *Suppose that Assumptions* 5.2.1, 5.2.2, *and* 5.3.1 *are satisfied. Then the necessary conditions for* ΔJ *to be nonnegative to first order for strong perturbations in the control* (5.26) *are that*

$$\dot{x}^o(t) = A(t)x^o(t) + B(t)u^o(t), \quad x^o(t_0) = x_0, \tag{5.41}$$

$$\dot{\lambda}(t) = -A^T(t)\lambda(t) - Q(t)x^o(t) - C^T(t)u^o(t), \tag{5.42}$$

$$\lambda(t_f) = S_f x^o(t_f) + D^T\nu, \tag{5.43}$$

$$0 = Dx^o(t), \tag{5.44}$$

$$0 = R(t)u^o(t) + C(t)x^o(t) + B^T(t)\lambda(t). \tag{5.45}$$

The remaining difficulty resides with the convexity of the cost associated with the neglected second-order terms. The objective of the remaining sections is to give necessary and sufficient conditions for optimality and to understand more deeply the character of the optimal solution when it exists. We have derived first-order necessary conditions for the minimization of the quadratic cost criterion. These necessary conditions form a two-point boundary-value problem in which the boundaries are linearly related through transition matrices. It will be shown that the Lagrange multipliers are linearly related to the state variables. Furthermore, the optimal cost criterion is shown to be represented by a quadratic function of the state. This form is reminiscent of the optimal value function used in the H-J-B theory of Chapter 3 to solve the LQ problem, and here we show how to extend the LQ problem to terminal constraints and free terminal time for the H-J-B theory given in Chapter 4. In this way we relate the Lagrange multiplier to the derivative of the optimal value function with respect to the state as given explicitly in Section 3.5 for terminally unconstrained optimization problems. This generalization of the optimal value function to include terminal constraints and free terminal time implies that the optimal solution to this general formulation of the LQ problem is not just a local minimum but a global minimum.

5.4 LQ Problem without Terminal Constraints: Transition Matrix Approach

By applying the first-order necessary conditions of Theorem 5.3.1 to the problem of Equations (5.17) through (5.19), the resulting necessary conditions for the unconstrained terminal problem are given by (5.41) to (5.45), where in (5.43) $D = 0$. By Assumption 5.2.1, the extremal $u^o(t)$ can be determined from (5.45) as a function of $x^o(t)$ and $\lambda(t)$ as

$$u^o(t) = -R^{-1}(t)\left[C(t)x^o(t) + B^T(t)\lambda(t)\right]. \tag{5.46}$$

Substitution of $u^o(t)$ given by (5.46) into (5.41) and (5.42) results in the linear homogeneous $2n$-vector differential equation

$$\begin{bmatrix} \dot{x}^o(t) \\ \dot{\lambda}(t) \end{bmatrix} = \begin{bmatrix} A(t) - B(t)R^{-1}(t)C(t) & -B(t)R^{-1}(t)B^T(t) \\ -Q(t) + C^T(t)R^{-1}(t)C(t) & -(A(t) - B(t)R^{-1}(t)C(t))^T \end{bmatrix} \begin{bmatrix} x^o(t) \\ \lambda(t) \end{bmatrix}. \tag{5.47}$$

Equation (5.47) is solved as a two-point boundary-value problem where n conditions in (5.41) are given at the initial time and n conditions are specified at the final time as $\lambda(t_f) = S_f x^o(t_f)$, i.e., (5.43) with $D = 0$.

For convenience, define the Hamiltonian matrix as

$$H(t) \triangleq \begin{bmatrix} A(t) - B(t)R^{-1}(t)C(t) & -B(t)R^{-1}(t)B^T(t) \\ -Q(t) + C^T(t)R^{-1}(t)C(t) & -(A(t) - B(t)R^{-1}(t)C(t))^T \end{bmatrix} \tag{5.48}$$

and the transition matrix (see Section A.3) associated with the solution of

$$\frac{d}{dt}\Phi_H(t,\tau) = \dot{\Phi}_H(t,\tau) = H(t)\Phi_H(t,\tau) \tag{5.49}$$

in block-partitioned form is

$$\Phi_H(t,\tau) \triangleq \begin{bmatrix} \Phi_{11}(t,\tau) & \Phi_{12}(t,\tau) \\ \Phi_{21}(t,\tau) & \Phi_{22}(t,\tau) \end{bmatrix}, \tag{5.50}$$

where t is the output (or solution) time and τ is the input (or initial) time. Using this block-partitioned transition matrix, the solution to (5.47) is represented as

$$\begin{bmatrix} x^o(t_f) \\ \lambda(t_f) \end{bmatrix} = \begin{bmatrix} x^o(t_f) \\ S_f x^o(t_f) \end{bmatrix} = \begin{bmatrix} \Phi_{11}(t_f,t_0) & \Phi_{12}(t_f,t_0) \\ \Phi_{21}(t_f,t_0) & \Phi_{22}(t_f,t_0) \end{bmatrix} \begin{bmatrix} x^o(t_0) \\ \lambda(t_0) \end{bmatrix}. \tag{5.51}$$

The objective is to obtain a unique relation between $\lambda(t_0)$ and $x^o(t_0)$. From (5.51), the first matrix equation gives

$$x^o(t_f) = \Phi_{11}(t_f,t_0)x^o(t_0) + \Phi_{12}(t_f,t_0)\lambda(t_0). \tag{5.52}$$

The second matrix equation of (5.51), using (5.52) to eliminate $x^o(t_f)$, becomes

$$S_f[\Phi_{11}(t_f,t_0)x^o(t_0) + \Phi_{12}(t_f,t_0)\lambda(t_0)] = \Phi_{21}(t_f,t_0)x^o(t_0) + \Phi_{22}(t_f,t_0)\lambda(t_0). \tag{5.53}$$

By solving for $\lambda(t_0)$, assuming the necessary matrix inverse exists,

$$\lambda(t_0) = [S_f\Phi_{12}(t_f,t_0) - \Phi_{22}(t_f,t_0)]^{-1}[\Phi_{21}(t_f,t_0) - S_f\Phi_{11}(t_f,t_0)]x^o(t_0). \tag{5.54}$$

The invertibility of this matrix is crucial to the problem and is discussed in detail in the following sections. The result (5.54) is of central importance to the LQ theory because this allows the optimal control (5.46) to be expressed as an explicit linear function of the state. For convenience, define

$$S(t_f, t; S_f) \triangleq [S_f \Phi_{12}(t_f, t) - \Phi_{22}(t_f, t)]^{-1}[\Phi_{21}(t_f, t) - S_f \Phi_{11}(t_f, t)]. \tag{5.55}$$

If the transition matrix is evaluated over the interval $[t_0, t]$, where $t_0 \le t \le t_f$, then

$$\begin{bmatrix} x^o(t) \\ \lambda(t) \end{bmatrix} = \Phi_H(t, t_0) \begin{bmatrix} I \\ S(t_f, t_0; S_f) \end{bmatrix} x^o(t_0). \tag{5.56}$$

Substitution of (5.56) into (5.46) results in the general optimal control rule

$$u^o(t) = -R^{-1}(t) \left[C(t) \; B^T(t)\right] \Phi_H(t, t_0) \begin{bmatrix} I \\ S(t_f, t_0; S_f) \end{bmatrix} x^o(t_0). \tag{5.57}$$

This control rule can be interpreted as a sampled data controller if $x^o(t_0)$ is the state measured at the last sample time t_0 and t, the present time, lies within the interval $[t_0, t_0 + \Delta]$, where Δ is the time between samples. If t_0 is considered to be the present time, t, then (5.57) reduces to the optimal control rule

$$u^o(t) = -R(t)^{-1}[C(t) + B^T(t) S(t_f, t; S_f)] x^o(t). \tag{5.58}$$

This linear control rule forms the basis for LQ control synthesis.

Our objective is to understand the properties of this control rule. More precisely, the character of $S(t_f, t; S_f)$ and the transition matrix $\Phi_H(t, t_0)$ are to be studied. Beginning in the next subsection, some properties peculiar to Hamiltonian systems are presented.

5.4.1 Symplectic Properties of the Transition Matrix of Hamiltonian Systems

The transition matrix of the Hamiltonian system [37, 46] has the useful symplectic property, defined as

$$\Phi_H(t, t_0) \underline{J} \Phi_H^T(t, t_0) = \underline{J}, \tag{5.59}$$

where

$$\underline{J} \triangleq \begin{bmatrix} 0 & I \\ -I & 0 \end{bmatrix} \tag{5.60}$$

is known as the fundamental symplectic matrix.

To show that the transition matrix of our Hamiltonian system satisfies (5.59), we time-differentiate (5.59) as

$$\dot{\Phi}_H(t,t_0)\underline{J}\Phi_H^T(t,t_0) + \Phi_H(t,t_0)\underline{J}\dot{\Phi}_H^T(t,t_0) = 0. \tag{5.61}$$

Substitution of the differential equation for $\Phi_H(t,t_0)$ (5.49) into (5.61) results in

$$H(t)\Phi_H(t,t_0)\underline{J}\Phi_H^T(t,t_0) + \Phi_H(t,t_0)\underline{J}\Phi_H^T(t,t_0)H^T(t) = 0. \tag{5.62}$$

By using (5.59), (5.62) reduces to

$$H(t)\underline{J} + \underline{J}H^T(t) = 0, \tag{5.63}$$

where H(t) defined by (5.48) clearly satisfies (5.63). To obtain (5.59) by starting with the transpose of (5.63), multiplying it on the left by $\Phi_H(t,t_0)$ and on the right by $\Phi_H^T(t,t_0)$, and then substituting in (5.49), we obtain the exact differential

$$\frac{d}{dt}\Phi_H(t,t_0)\underline{J}^T\Phi_H^T(t,t_0) = 0. \tag{5.64}$$

The integral equals a constant matrix which when evaluated at t_0 is $\underline{J}^T$. This results in the form

$$\Phi_H^T(t,t_0)\underline{J}^T\Phi_H(t,t_0) = \underline{J}^T. \tag{5.65}$$

Taking the transpose of (5.65), multiplying it on the left by $-\Phi_H(t,t_0)\underline{J}$ and on the right by $-\Phi_H^{-1}(t,t_0)\underline{J}$, and using $\underline{J}\underline{J} = -I$, we obtain (5.59).

We now consider the spectral properties of $\Phi_H(t,t_0)$. Note that since $\underline{J}^T = \underline{J}^{-1}$, then from (5.59)

$$\Phi_H^{-1}(t,t_0) = \underline{J}^T \Phi_H^T(t,t_0)\underline{J}. \tag{5.66}$$

Furthermore, the characteristic equations of $\Phi_H(t,t_0)$ and $\Phi_H(t,t_0)^{-1}$ are the same since

$$\begin{aligned} \det(\Phi_H^{-1}(t,t_0) - \lambda I) &= \det(\underline{J}^T \Phi_H^T(t,t_0)\underline{J} - \lambda I) \\ &= \det(\underline{J}^T \Phi_H^T(t,t_0)\underline{J} - \lambda \underline{J}^T \underline{J}) \\ &= \det(\Phi_H(t,t_0) - \lambda I). \end{aligned} \tag{5.67}$$

The implication of this is that since $\Phi_H(t,t_0)$ is a $2n \times 2n$ nonsingular matrix, if μ_i, $i = 1, \dots, n$, are n eigenvalues of $\Phi_H(t,t_0)$ with $\mu_i \neq 0$ for all i, then the remaining n eigenvalues are

$$\mu_{i+n} = \frac{1}{\mu_i}, \quad i = 1, \dots, n. \tag{5.68}$$

By partitioning $\Phi_H(t,t_0)$ as in (5.50), the following relations are obtained from (5.59):

$$\Phi_{11}(t,t_0)\Phi_{12}^T(t,t_0) = \Phi_{12}(t,t_0)\Phi_{11}^T(t,t_0), \tag{5.69}$$

$$\Phi_{21}(t,t_0)\Phi_{22}^T(t,t_0) = \Phi_{22}(t,t_0)\Phi_{21}^T(t,t_0), \tag{5.70}$$

$$\Phi_{11}(t,t_0)\Phi_{22}^T(t,t_0) - \Phi_{12}(t,t_0)\Phi_{21}^T(t,t_0) = I. \tag{5.71}$$

These identities will be used later.

5.4.2 Riccati Matrix Differential Equation

Instead of forming $S(t_f,t;S_f)$ in (5.55) by calculating the transition matrix, in this section we show that $S(t_f,t;S_f)$ can be propagated directly by a quadratic matrix differential equation called the matrix Riccati equation. This equation plays a key role in all the analyses of the following sections.

A few preliminaries will help simplify this derivation as well as others in the coming sections. First, note that since $\Phi_H(t,\tau)$ is symplectic, then

$$\bar{\Phi}_H(t,\tau) \triangleq L\Phi_H(t,\tau) \tag{5.72}$$

with the symplectic matrix

$$L \triangleq \begin{bmatrix} I & 0 \\ -S_f & I \end{bmatrix} \tag{5.73}$$

is also a symplectic matrix. Therefore, the partitioned form of $\bar{\Phi}_H(t,\tau)$ satisfies (5.69) to (5.71). Second, by using the propagation equation for the transition matrix (5.49), differentiation of the identity

$$\Phi_H(t_f,t)\Phi_H(t,t_0) = \Phi_H(t_f,t_0) \tag{5.74}$$

with respect to t, where t_f and t_0 are two fixed times, gives the adjoint form for propagating the transition matrix as

$$\frac{d}{dt}\Phi_H(t_f,t) = -\Phi_H(t_f,t)H(t), \quad \Phi_H(t_f,t_f) = I, \tag{5.75}$$

where input time is the independent variable and the output time is fixed. Therefore,

$$\frac{d}{dt}\bar{\Phi}_H(t_f,t) = -\bar{\Phi}_H(t_f,t)H(t), \quad \bar{\Phi}_H(t_f,t_f) = L. \tag{5.76}$$

Finally, note from (5.55) and (5.72) that

$$S(t_f,t;S_f) = -\bar{\Phi}_{22}^{-1}(t_f,t)\bar{\Phi}_{21}(t_f,t). \tag{5.77}$$

Theorem 5.4.1 *Let $\Phi_H(t_f, t)$ be the transition matrix for the dynamic system* (5.47) *with partitioning given by* (5.50). *Then a symmetric matrix $S(t_f, t; S_f)$ satisfies*

$$\begin{aligned} \frac{d}{dt}S(t_f, t; S_f) &= -\left[A(t) - B(t)R^{-1}(t)C(t)\right]^T S(t_f, t; S_f) \\ &\quad - S(t_f, t; S_f)\left[A(t) - B(t)R^{-1}(t)C(t)\right] \\ &\quad - \left[Q(t) - C^T(t)R^{-1}(t)C(t)\right] \\ &\quad + S(t_f, t; S_f)B(t)R^{-1}(t)B^T(t)S(t_f, t; S_f), \\ &\quad S(t_f, t_f; S_f) = S_f \end{aligned} \tag{5.78}$$

if the inverse in (5.77) *exists and S_f is symmetric.*

Proof: If (5.77) is rewritten as

$$-\bar{\Phi}_{21}(t_f, t) = \bar{\Phi}_{22}(t_f, t)S(t_f, t; S_f) \tag{5.79}$$

and differentiated with respect to t, then

$$\begin{aligned} -\frac{d}{dt}\left[\bar{\Phi}_{21}(t_f, t)\right] &= \frac{d}{dt}\left[\bar{\Phi}_{22}(t_f, t)\right] S(t_f, f; S_f) \\ &\quad + \left[\bar{\Phi}_{22}(t_f, t)\right]\frac{d}{dt}\left[S(t_f, f; S_f)\right]. \end{aligned} \tag{5.80}$$

The derivatives for $\bar{\Phi}_{21}(t_f, t)$ and $\bar{\Phi}_{22}(t_f, t)$ are obtained from the partitioning of (5.76) as

$$\begin{aligned} &\bar{\Phi}_{21}(t_f, t)(A(t) - B(t)R^{-1}(t)C(t)) - \bar{\Phi}_{22}(t_f, t)(Q(t) - C^T(t)R^{-1}(t)C(t)) \\ &= \left[\bar{\Phi}_{21}(t_f, t)B(t)R^{-1}(t)B^T(t) + \bar{\Phi}_{22}(t_f, t)(A(t)\right. \\ &\quad \left.-B(t)R^{-1}(t)C(t))^T\right] S(t_f, t; S_f) + \bar{\Phi}_{22}(t_f, t)\frac{dS(t_f; S_f)}{dt}. \end{aligned} \tag{5.81}$$

Since $\bar{\Phi}_{22}(t_f, t)$ is assumed invertible, then by premultiplying by $\bar{\Phi}_{22}(t_f, t)^{-1}$ and using (5.77), the Riccati equation of (5.78) is obtained. Since S_f is symmetric, and $\frac{d}{dt}S(t_f, t; S_f) \triangleq \dot{S}(t_f, t; S_f) = \dot{S}^T(t_f, t; S_f)$, then $S(t_f, t; S_f)$ will be symmetric. ∎

Remark 5.4.1 *By the symplectic property* (5.70), *using* (5.77),

$$S(t_f, t; S_f) = -\bar{\Phi}_{22}^{-1}(t_f, t)\bar{\Phi}_{21}(t_f, t) = -\bar{\Phi}_{21}^{T}(t_f, t)\bar{\Phi}_{22}^{-T}(t_f, t) \tag{5.82}$$

implies that $S(t_f, t; S_f)$ *is symmetric, without directly using the Riccati differential equation* (5.78).

Remark 5.4.2 *Since the boundary condition on* $\bar{\Phi}_{22}(t_f, t_f)$ *is the identity matrix, a finite interval of time is needed before* $\bar{\Phi}_{22}(t_f, t)$ *may no longer be invertible and the control law of* (5.58) *would no longer be meaningful. In the classical calculus of variations literature this is the focal point condition or Jacobi condition* [22].

In the next section, it is shown that if the Riccati variable $S(t_f, t; S_f)$ exists, then the Hamiltonian system can be transformed into another similar Hamiltonian system for which the feedback law appears directly in the dynamic system.

5.4.3 Canonical Transformation of the Hamiltonian System

Some additional insight into the character of this Hamiltonian system is obtained by a canonical similarity transformation of the variables $x(t)$ and $\lambda(t)$ into a new set of variables $x^o(t)$ and $\bar{\lambda}(t)$. A transformation $L(t)$ for Hamiltonian systems is said to be canonical if it satisfies the symplectic property

$$L(t)\underline{J}L^T(t) = \underline{J}, \tag{5.83}$$

where $\underline{J}$ is defined in (5.60).

The canonical transformation that produces the desired result is

$$L(t) \triangleq \begin{bmatrix} I & 0 \\ -S(t_f, t; S_f) & I \end{bmatrix} \tag{5.84}$$

such that

$$\begin{bmatrix} x^o(t) \\ \bar{\lambda}(t) \end{bmatrix} \triangleq \begin{bmatrix} I & 0 \\ -S(t_f, t; S_f) & I \end{bmatrix} \begin{bmatrix} x^o(t) \\ \lambda(t) \end{bmatrix} \tag{5.85}$$

$L(t)$ is a canonical transformation since $S(t_f, t; S_f)$ is symmetric. Note that the state variables are not being transformed. The propagation equation for the new variables is obtained by differentiating (5.85) and using (5.47) and (5.78) as

$$\begin{aligned}\begin{bmatrix} \dot{x}^o(t) \\ \dot{\bar{\lambda}}(t) \end{bmatrix} &= \begin{bmatrix} A - BR^{-1}(C + B^T S) & -BR^{-1}B^T \\ 0 & -\left[A - BR^{-1}(C + B^T S)\right]^T \end{bmatrix} \begin{bmatrix} x^o(t) \\ \bar{\lambda}(t) \end{bmatrix}, \\ \begin{bmatrix} x^o(t_f) \\ \bar{\lambda}(t_f) \end{bmatrix} &= L(t_f) \begin{bmatrix} x^o(t_f) \\ S_f x^o(t_f) \end{bmatrix},\end{aligned} \tag{5.86}$$

where use is made of the inverse of $L(t)$:

$$L^{-1}(t) = \begin{bmatrix} I & 0 \\ S(t_f, t; S_f) & I \end{bmatrix}. \tag{5.87}$$

The zero matrix in the coefficient matrix of (5.83) is a direct consequence of $S(t_f, t; S_f)$ satisfying the Riccati equation (5.78)

Note that from the boundary condition of (5.86)

$$\bar{\lambda}(t_f) = [S_f - S(t_f, t_f; S_f)]x^o(t_f). \tag{5.88}$$

However, since $S(t_f, t_f; S_f) = S_f$,

$$\bar{\lambda}(t_f) = 0. \tag{5.89}$$

Observe in (5.86) that $\bar{\lambda}(t)$ is propagated by a homogeneous differential equation, which, by (5.89), has the trivial solution $\bar{\lambda}(t) = 0$ for all t in the interval $[t_0, t_f]$. Therefore, (5.86) produces the differential equation for the state as

$$\begin{aligned}\dot{x}^o(t) &= [A(t) - B(t)R^{-1}(t)(C(t) + B^T(t)S(t_f, t, S_f))]x^o(t), \\ x^o(t_0) &= x_0,\end{aligned} \tag{5.90}$$

which is the dynamic equation using the optimal control rule (5.58).

The existence of the Riccati variable is necessary for the existence of the linear control rule and the above canonical transformation. In the next section it is shown that the existence of the Riccati variable $S(t_f, t; S_f)$ is a necessary and sufficient condition for the quadratic cost criterion to be positive definite, i.e., it is a necessary and sufficient condition for the linear controller to be the minimizing controller.

5.4.4 Necessary and Sufficient Condition for the Positivity of the Quadratic Cost Criterion

We show here that the quadratic cost criterion is actually positive for all controls which are not null when the initial condition is $x(t_0) = 0$. Conditions obtained here are applicable to the second variation of Section 5.1 being positive definite for the optimization problem without terminal constraints and fixed terminal time. The essential assumption, besides $R(t) > 0$, is complete controllability. The main ideas in this section were given in [20] and [4].

Assumption 5.4.1 *The dynamic system* (5.18) *is controllable on* any *interval* $[t, t']$ *where* $t_0 \le t < t' \le t_f$ *(completely controllable). If the system is completely controllable, then a bounded control*

$$\bar{u}(t) = B^T(t)\Phi_A^T(t', t)W_A^{-1}(t', t)x(t') \tag{5.91}$$

always exists which transfers the state $x(t_0) = 0$ *to any desired state* $x(t')$ *at* $t = t'$ *where* $W_A(t', t_0)$*, the controllability Grammian matrix, is*

$$W_A(t', t_0) = \int_{t_0}^{t'} \Phi_A(t', t)B(t)B^T(t)\Phi_A^T(t', t)dt > 0 \tag{5.92}$$

and

$$\frac{d}{dt}\Phi_A(t, \sigma) = A(t)\Phi_A(t, \sigma), \quad \Phi(\sigma, \sigma) = I, \tag{5.93}$$

for all t' *in* $(t_0, t_f]$.

Definition 5.4.1 $J(u(\cdot); 0, t_0)$ *is said to be positive definite if for each* $u(\cdot)$ *in* $\mathcal{U}$, $u(\cdot) \ne 0$ *(null function),* $J(u(\cdot); 0, t_0) > 0$.

We show first that if for all t in $[t_0, t_f]$ there exists an $S(t_f, t; S_f)$ which satisfies (5.78), then $J(u(\cdot); 0, t_0)$ is positive definite. Consider adding to $J(u(\cdot); x(t_0), t_0)$ of (5.17) the identically zero quantity

$$\frac{1}{2}x^T(t_0)S(t_f, t_0; S_f)x(t_0) - \frac{1}{2}x^T(t_f)S(t_f, t_f; S_f)x(t_f) + \int_{t_0}^{t_f} \frac{1}{2}\frac{d}{dt}x^T(t)S(t_f, t; S_f)x(t)dt = 0. \tag{5.94}$$

This can be rewritten using the Riccati equation (5.78) and the dynamics (5.18) as

$$\begin{aligned} &\frac{1}{2}x^T(t_0)S(t_f, t_0; S_f)x(t_0) - \frac{1}{2}x^T(t_f)S_f x(t_f) \\ &+ \int_{t_0}^{t_f} \Big\{ -\frac{1}{2}x^T(t)\Big[Q(t) + S(t_f, t; S_f)A(t) + A^T(t)S(t_f, t, S_f) \\ &- \left(C(t) + B^T(t)S(t_f, t; S_f)\right)^T R^{-1}(t)\left(C(t) + B^T(t)S(t_f, t; S_f)\right)\Big]x(t) \\ &+ \frac{1}{2}x^T(t)\left[S(t_f, t; S_f)A(t) + A^T(t)S(t_f, t; S_f)\right]x(t) \\ &+ x^T(t)S(t_f, t; S_f)B(t)u(t)\Big\}dt = 0. \end{aligned} \tag{5.95}$$

If (5.95) is added to (5.17), the integrand can be manipulated into a perfect square and the cost takes the form

$$\begin{aligned} &J(u(\cdot); x(t_0), t_0) \\ &= \frac{1}{2}x^T(t_0)S(t_f, t_0; S_f)x(t_0) \\ &+ \frac{1}{2}\int_{t_0}^{t_f} \left[u(t) + R^{-1}(t)\left(C(t) + B^T(t)S(t_f, t; S_f)\right)x(t)\right]^T \\ &\times R(t)\left[u(t) + R^{-1}(t)\left(C(t) + B^T(t)S(t_f, t; S_f)\right)x(t)\right]dt. \end{aligned} \tag{5.96}$$

Therefore, the cost takes on its minimum value when $u(t)$ takes the form of the optimal controller (5.58). If $x(t_0)$ is zero, then the optimal control is the null control. Any other control $u(\cdot) \neq u^o(\cdot)$ will give a positive value to $J(u(\cdot); 0, t_0)$. This shows that a sufficient condition for $J(u(\cdot); 0, t_0) > 0$ is the existence of $S(t_f, t; S_f)$ over the interval $[t_0, t_f]$.

Now we consider the necessity of the existence of $S(t_f, t; S_f)$ over the interval $[t_0, t_f]$ for $J(u(\cdot); 0, t_0) > 0$. If the cost $J(u(\cdot); 0, t_0)$ is positive definite, then the necessity of the existence of the Riccati variable $S(t_f, t; S_f)$ depends on Assumption 5.4.1. First note from Remark 5.4.2 that for some t' close enough to t_f, $S(t_f, t; S_f)$ exists for $t' \le t \le t_f$. Therefore, by using the optimal control from t' to t_f for some $x(t')$, the minimizing cost is

$$J(u^o(\cdot); x(t'), t') = \frac{1}{2} x^T(t') S(t_f, t'; S_f) x(t'). \tag{5.97}$$

We prove the necessity of the existence of $S(t_f, t; S_f)$ by supposing the opposite—that $S(t_f, t; S_f)$ ceases to exist at some escape time,[8] t_e. If this occurs, then we will show that the cost criterion can be made either as negative as we like, which violates the positive-definite assumption of the cost criterion, or as positive as desired, which violates the assumption of minimality. The cost, using the control defined as

$$\begin{aligned} u(t) &= \bar{u}(t), \quad t_0 \le t \le t', \\ u(t) &= u^o(t), \quad t' < t \le t_f, \end{aligned} \tag{5.98}$$

can be written as

$$\begin{aligned} J(u(\cdot); 0, t_0) &= \frac{1}{2} x^T(t') S(t_f, t'; S_f) x(t') \\ &+ \int_{t_0}^{t'} \left[\frac{1}{2} x^T(t) Q(t) x(t) + \bar{u}^T(t) C(t) x(t) + \frac{1}{2} \bar{u}^T(t) R(t) \bar{u}(t) \right] dt, \end{aligned} \tag{5.99}$$

where by Assumption 5.4.1 there exists a control $\bar{u}(t)$ for $t \in [t_0, t']$ which transfers $x(t_0) = 0$ to any $x(t')$ at $t' > t_0$ such that $\|x(t')\| = 1$ and $\|\bar{u}(\cdot)\| \le \rho(t') < \infty$, where $\rho(\cdot)$ is a continuous positive function.

[8]Since the Riccati equation is nonlinear, the solution may approach infinite values for finite values of time. This is called the escape time.

Since $\|\bar{u}(\cdot)\|$ is bounded by the controllability assumption, the integral in (5.99) is also bounded. If $x^T(t')S(t_f, t'; S_f)x(t') \to -\infty$ as $t' \to t_e$ $(t_e < t')$ for some $x(t')$, then $J(u(\cdot); 0, t_0) \to -\infty$. But this violates the assumption that $J(u(\cdot); 0, t_0) > 0$. Furthermore, $x(t')^T S(t_f, t'; S_f)x(t')$ cannot go to positive infinity since this implies that the minimal cost $J(u^o(\cdot); x(t'), t')$ can be made infinite: the controllability assumption implies that there exists a finite control which gives a finite cost. Therefore, since the integral in (5.99) can be bounded, $x^T(t')S(t_f, t'; S_f)x(t')$ cannot go to ∞. Our arguments apply to the interval $(t_0, t_f]$. We appeal to [22] to verify that $S(t_f, t; S_f)$ exists for all t in $[t_0, t_f]$. These results are summarized in the following theorem.

Theorem 5.4.2 *Suppose that system* (5.18) *is completely controllable. A necessary and sufficient condition for* $J(u(\cdot); 0, t_0)$ *to be positive definite is that there exists a function* $S(t_f, t; S_f)$ *for all* t *in* $[t_0, t_f]$ *which satisfies the Riccati equation* (5.78).

Note that since no smallness requirements are placed on $x(t)$ as in Section 5.4, Theorem 5.4.2 is a statement of the global optimality of the solution to the LQ problem. This was shown in Example 3.5.1, where the optimal value function used in the Hilbert's integral was a quadratic form identical to the function given in (5.97).

Remark 5.4.3 *The optimal value function, given by* (5.97), *is*

$$V(x(t), t) = \frac{1}{2} x^T(t) S(t_f, t; S_f) x(t) \tag{5.100}$$

and is valid as long as the solution $S(t_f, t; S_f)$ *to the Riccati equation* (5.78) *exists. This quadratic form satisfies from Theorem* 3.5.2 *the Hamilton–Jacobi–Bellman equation, again showing that the solution to the LQ problem is global. Furthermore,* $V_x^T(x(t), t) = S(t_f, t; S_f)x(t)$ *satisfies the same differential equation as* $\lambda(t)$ *and* $V_{xx}(x(t), t) = S(t_f, t; S_f)$ *satisfies a Riccati equation as given in Section* 3.5.1.

5.4.5 Necessary and Sufficient Conditions for Strong Positivity

The positivity of the second variation as represented by $J(u(\cdot); 0, t_0)$ is not enough to ensure that the second variation dominates the higher-order term in the expansion of the cost criterion about the assumed minimizing path. To ensure that the second variation dominates the expansion, it is sufficient that the second variation be strongly positive [22]. Although the LQ problem does not need strong positivity, in the second variational problem if the second-order term is to dominate over higher-order terms in the expansion (see (5.8)), strong positivity is required. In this section we prove that a necessary and sufficient condition for strong positivity of the nonsingular ($R = H_{uu} > 0$) second variation is that a solution exists to the matrix Riccati differential equation (5.78). The sufficiency part of this theorem is very well known and documented [8]. Though the necessity part is well known too, it is not, in our opinion, proved convincingly elsewhere except in certain special cases.

Strong positivity is defined as follows.

Definition 5.4.2 *$J(u(\cdot); 0, t_0)$ is said to be strongly positive if for each $u(\cdot)$ in $\mathcal{U}$, and some $k > 0$,*

$$J(u(\cdot); 0, t_0) \geq k \|u(\cdot)\|^2, \tag{5.101}$$

where $\|u(\cdot)\|$ is some suitable norm defined on $\mathcal{U}$.

The extension from showing that the second variation is positive definite (Theorem 5.4.2) to strongly positive is given in the following theorem.

Theorem 5.4.3 *A necessary and sufficient condition for $J(u(\cdot); 0, t_0)$ to be strongly positive is that for all t in $[t_0, t_f]$ there exists a function $S(t_f, t; S_f)$ which satisfies the Riccati equation (5.78).*

Proof: In Theorem 5.4.2 we proved that $J\left(u(\cdot);0,t_0\right)$ is positive definite. To show that $J\left(u(\cdot);0,t_0\right)$ is strongly positive, consider a new LQ problem with cost criterion

$$\begin{aligned} &J\left(u(\cdot),\varepsilon;0,t_0\right) \\ &= \int_{t_0}^{t_f}\left[\frac{1}{2}x^T(t)Q(t)x(t)+u^T(t)C(t)x(t)+\frac{1}{2-\varepsilon}u^T(t)R(t)u(t)\right]dt \\ &+\frac{1}{2}x^T(t_f)S_fx(t_f). \end{aligned} \tag{5.102}$$

This functional is positive definite if $2-\varepsilon>0$ and if and only if

$$\begin{aligned} -\dot{S}(t_f,t;S_f,\varepsilon) &= Q(t)+S(t_f,t;S_f,\varepsilon)A(t)+A^T(t)S(t_f,t;S_f,\varepsilon) \\ &\quad -[C(t)+B^T(t)S(t_f,t;S_f,\varepsilon)]^T\frac{2-\varepsilon}{2}R^{-1}(t)[C(t) \\ &\quad +B^T(t)S(t_f,t;S_f,\varepsilon)], \end{aligned} \tag{5.103}$$

$$S(t_f,t_f;S_f,\varepsilon) = S_f, \tag{5.104}$$

has a solution $S(t_f,t;S_f,\varepsilon)$ defined for all t in $[t_0,t_f]$.

Now, since $Q(t)$, $C(t)$, $R(t)$, $A(t)$, $B(t)$ are continuous in t and the right-hand side of (5.103) is analytic in $S(t_f,t;S_f,\varepsilon)$ and ε, and since $S(t_f,\cdot;S_f,0)$ exists, we have that $S(t_f,\cdot;S_f,\varepsilon)$ is a continuous function of ε at $\varepsilon=0$ [15]. So, for ε sufficiently small, $S(t_f,t;S_f,\varepsilon)$ exists for all t in $[t_0,t_f]$. Therefore, for $\varepsilon<0$ and sufficiently small, $J\left(u(\cdot),\varepsilon;0,t_0\right)$ is positive definite.

Next, we note that

$$J\left(u(\cdot),\varepsilon;0,t_0\right)=J\left(u(\cdot);0,t_0\right)+\frac{\varepsilon}{2(2-\varepsilon)}\int_{t_0}^{t_f}u^T(t)R(t)u(t)dt\geq 0 \tag{5.105}$$

so that

$$J\left(u(\cdot);0,t_0\right)\geq-\frac{\varepsilon}{2(2-\varepsilon)}\int_{t_0}^{t_f}u^T(t)R(t)u(t)dt. \tag{5.106}$$

From Assumption 5.2.1, $R(t) > 0$ with norm bound $0 < k_1 \leq ||R(t)|| \leq k_2$. Therefore, we conclude from (5.106) that

$$J\left(u(\cdot);0,t_0\right) \geq -\frac{\varepsilon k_1}{2(2-\varepsilon)}\int_{t_0}^{t_f} u^T(t)u(t)dt, \quad k_1 > 0. \tag{5.107}$$

Hence,[9]

$$J\left(u(\cdot);0,t_0\right) \geq k\|u(\cdot)\|_{\mathcal{L}_2}^2, \tag{5.108}$$

where

$$k = -\frac{\varepsilon k_1}{2(2-\varepsilon)} > 0. \tag{5.109}$$

which implies that $J\left(u(\cdot);0,t_0\right)$ is strongly positive.

The converse is found in the proof of Theorem 5.4.2, namely, that if $J\left(u(\cdot);0,t_0\right)$ is strongly positive, then an $S(S_f,\cdot,S_f)$ exists which satisfies (5.78). ▮

Remark 5.4.4 *For $J\left(u(\cdot);0,t_0\right)$ strongly positive the second variation dominates all higher-order terms in the expansion of the cost criterion about the minimizing path with fixed terminal time and without terminal constraints. See Section* 5.1.

Example 5.4.1 (Shortest distance between a point and a great circle) *This example illustrates that the second variation is no longer positive definite when the solution to the Riccati equation* (5.78) *escapes and that other neighboring paths can produce smaller values of the cost. Let s be the distance along a path on the surface of a sphere. The differential distance is*

$$ds = (r^2 d\theta^2 + r^2\cos^2\theta d\phi^2)^{\frac{1}{2}} = r^2(u^2+\cos^2\theta)^{\frac{1}{2}}d\phi, \tag{5.110}$$

[9] $\int_{t_0}^{t_f} u^T(t)u(t)dt = \|u(\cdot)\|_{\mathcal{L}_2}^2$ is the $\mathcal{L}_2$ or integral square norm.

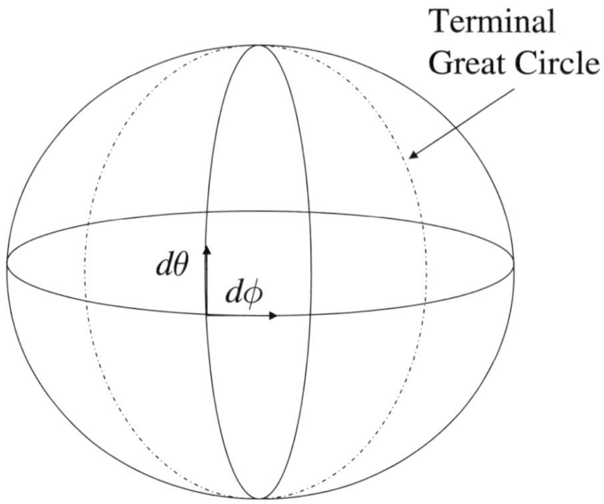

Figure 5.1: Coordinate frame on a sphere.

where ϕ *is the longitudinal angle and* θ *is the lateral angular as shown in Figure* 5.1. *The dynamic equation is*

$$\frac{d\theta}{d\phi} = u, \ \theta(0) = 0, \tag{5.111}$$

where u *is the control variable and* ϕ *is treated as the independent variable. The problem is to minimize the distance on a unit sphere* ($r = 1$) *from a given point to a given great circle, i.e., find the control* u *that minimizes*

$$J\left(u(\cdot); 0, \phi = 0\right) = \int_0^{\phi_1} (u^2 + \cos^2\theta)^{\frac{1}{2}} d\phi \tag{5.112}$$

subject to (5.111), *fixed terminal* $\phi = \phi_1$, *and unconstrained* $\theta(\phi_1)$.

First, a trajectory that satisfies the first-order necessary conditions is determined. Then, about that path, the second-order necessary conditions will be analyzed using the conditions developed in Section 5.4.4. *The variational Hamiltonian is*

$$H = (u^2 + \cos^2\theta)^{\frac{1}{2}} + \lambda u. \tag{5.113}$$

The first-order necessary conditions are

$$\frac{d\theta}{d\phi} = H_\lambda = u, \ \theta(0) = 0, \tag{5.114}$$

$$-\dot{\lambda} = H_\theta = -(u^2 + \cos^2\theta)^{-\frac{1}{2}} \sin\theta \cos\theta, \ \lambda(\phi_1) = 0, \tag{5.115}$$

$$H_u = \lambda + u(u^2 + \cos^2\theta)^{-\frac{1}{2}} = 0. \tag{5.116}$$

The first-order necessary conditions are satisfied by the trajectory $u(\phi) = 0$, $\lambda(\phi) = 0$, $\theta(\phi) = 0$ *for* $0 \le \phi \le \phi_1$.

About this extremal path, the second-order necessary conditions are generated. The second variational problem given in Section 5.2 *is to find the perturbed control* u *that minimizes the second variational cost criterion*

$$\delta^2 J = \frac{1}{2} \int_0^{\phi_1} \left[(\delta u)^2 - (\delta\theta)^2 \right] d\phi \tag{5.117}$$

subject to $\frac{d\delta\theta}{d\phi} = \delta u$, $\delta\theta(0) = 0$ *for fixed terminal* $\phi = \phi_1$ *with no constraint on the terminal value of* $\theta(\phi_1)$. *From Theorem* 5.4.2 *it is necessary that the solution of the associated Riccati equation* (5.78) *exist. Since for this example* $A = 0$, $B = 1$, $C = 0$, $Q = -1$, $R = 1$, *the associated Riccati equation and solution are*

$$\frac{dS}{d\phi} = S^2 + 1, \ S(\phi_1) = 0 \Rightarrow S(\phi) = -\tan(\phi_1 - \phi). \tag{5.118}$$

Note that the solution remains finite until it escapes at $\phi_1 - \phi = \pi/2$. *At that point any great circle path from the point to the terminal great circle will give the same cost. If the path is longer than* $\pi/2$, *then there are neighboring paths that do not even satisfy the first-order necessary conditions, which can give smaller cost. If* $\phi_1 - \phi < \pi/2$, *then the second variation is positive and can be shown to be strongly positive. Furthermore, the second variational controller is* $\delta u = \tan(\phi_1 - \phi)\delta\theta$. *Note that the gain is positive indicating that the best neighboring optimum controller is essentially divergent.*

5.4.6 Strong Positivity and the Totally Singular Second Variation

Unfortunately, it turns out that if $R(t) = H_{uu} \ge 0$, the so-called singular control problem, the second variation cannot be strongly positive [4] and so different tests for sufficiency have to be devised. In the case of nonsingular optimal control problems, where $R(t) = H_{uu} > 0$ is invertible for all t in $[t_0, t_f]$, it is known that a sufficient condition for strong positivity of the second variation, and hence for a weak local minimum, is that the matrix Riccati differential equation associated with the second variation should have a solution for all t in $[t_0, t_f]$. Clearly, this condition is inapplicable in the singular case owing to the presence of $R^{-1}(t) = H_{uu}^{-1}$ in the matrix Riccati equation. For a long time, therefore, it was felt that no Riccati-like condition existed for the singular case. This has turned out to be not true [4]. The result is that sufficiency conditions for nonnegativity of singular and nonsingular second variations are rather closely related.

In Section 5.4.5 we demonstrated that if the matrix Riccati equation has a solution for all t in $[t_0, t_f]$, then $J(u(\cdot); 0, t_0)$ is not only positive definite but also strongly positive. Clearly, in a finite-dimensional vector space, positive definiteness is equivalent to strong positivity. However, in our space of piecewise continuous control functions this is not so. In this section we illustrate the difference between positive definiteness and strong positivity by means of a simple example. In addition, the example illustrates the fact that the totally singular second variation cannot be strongly positive. (See [4] for general proofs.) This is, of course, consistent with the fact that in the totally singular case the matrix Riccati equation is undefined because $R^{-1}(t)$ does not exist.

Before presenting the example, we define "totally singular" precisely as follows.

Definition 5.4.3 $J(u(\cdot); 0, t_0)$ *is said to be totally singular if*

$$R(t) = 0 \quad \forall\, t \ \text{in} \ [t_0, t_f]. \tag{5.119}$$

Now, we consider the totally singular functional

$$J(u(\cdot); 0, t_0) = \int_{t_0}^{t_f} x^2(t)dt \tag{5.120}$$

subject to

$$\dot{x}(t) = u(t), \quad x(t_0) = 0, \tag{5.121}$$

$$u(\cdot) \text{ is a member of } \mathcal{U}. \tag{5.122}$$

Clearly $J\left(u(\cdot); 0, t_0\right)$ is positive definite.

Set

$$u(t) = \cos \omega(t - t_0), \quad t_0 \le t \le t_f, \tag{5.123}$$

so that

$$x(t) = \frac{1}{\omega} \sin \omega(t - t_0), \quad t_0 \le t \le t_f. \tag{5.124}$$

With this choice of control $J\left(u(\cdot); 0, t_0\right)$ becomes

$$J\left(u(\cdot); 0, t_0\right) = \int_{t_0}^{t_f} \frac{1}{\omega^2} \sin^2 \omega(t - t_0)\, dt, \tag{5.125}$$

and

$$\|u(\cdot)\|_{\mathcal{L}_2}^2 = \int_{t_0}^{t_f} u^2(t) dt = \int_{t_0}^{t_f} \cos^2 \omega(t - t_0) dt = \int_{t_0}^{t_f} (1 - \sin^2 \omega(t - t_0)) dt. \tag{5.126}$$

By definition, if $J\left(u(\cdot); 0, t_0\right)$ were strongly positive, then for some $k > 0$ and all $\omega > 0$,

$$\frac{1}{\omega^2} \int_{t_0}^{t_f} (\sin^2 \omega(t - t_0)) dt \ge k \int_{t_0}^{t_f} u^2(t) dt = k \int_{t_0}^{t_f} (1 - \sin^2 \omega(t - t_0)) dt, \tag{5.127}$$

i.e.,

$$\left(k + \frac{1}{\omega^2}\right) \int_{t_0}^{t_f} (\sin^2 \omega(t - t_0)) dt \ge k(t_f - t_0). \tag{5.128}$$

But this is impossible because the left-hand side of (5.128) tends to $\frac{1}{2}k(t_f - t_0)$ as $\omega \to \infty$. In other words, $J\left(u(\cdot); 0, t_0\right)$ of (5.120) is not strongly positive.

Note, however, that the functional

$$J\left(u(\cdot);0,t_0\right)=\int_{t_0}^{t_f}(x^2(t)+u^2(t))dt \tag{5.129}$$

is strongly positive. This follows directly because

$$x^2(\cdot)+u^2(\cdot)\geq u^2(\cdot). \tag{5.130}$$

The fact that the totally singular second variation cannot be strongly positive implies that in the totally singular case we should seek (necessary and) sufficient conditions only for nonnegativity and for positive definiteness of $J\left(u(\cdot);0,t_0\right)$. For a full treatment of the singular optimal control problem, see [4] and [14].

5.4.7 Solving the Two-Point Boundary-Value Problem via the Shooting Method

A second-order method for finding an optimal control numerically is with a shooting method. This method was contrasted with the first-order steepest descent method in Section 3.3.4 but can be described in detail here. To implement the shooting method, guess a value for the initial value of the adjoint vector, $\lambda_i(t_0)$. Integrate this guess forward along with the state to t_f using (3.41) and (3.56) with the optimal control determined from (3.55). Now, perturb the previous guess $\lambda_i(t_0)$ by some function of the errors at the terminal time t_f to get your next guess $\lambda_{i+1}(t_0)$. It should be noted that small variations in $\lambda(t_0)$ can lead to large changes in $\lambda(t)$, depending on the stability properties of the adjoint equation. State equations that are stable with time running forward imply that the associated adjoint equations are unstable when they are integrated forward in time (but stable when they are integrated *backward* in time) [11].

We follow the development in Section 5.1. Assume $f(x(t),u(t),t)$ and $L(x(t),u(t),t)$ are twice differentiable in $x(t)$ and $u(t)$. Assume also that $H_{uu}(x(t),u(t),\lambda(t),t)$ is nonsingular along all trial trajectories.

1. Choose $\lambda^i(t_0)$ where i is the iteration index. Define

$$y^i(t) = \begin{bmatrix} x^i(t) \\ \lambda^i(t) \end{bmatrix}. \tag{5.131}$$

The initial condition, $x(t_0) = x_0$, is given. The control is calculated from

$$H_u(x^i(t), u^i(t), \lambda^i(t), t) = 0 \Rightarrow u^i(t) = g(x^i(t), \lambda^i(t), t) \tag{5.132}$$

from the Implicit Function Theorem. This trajectory *does not* satisfy the given $\lambda(t_f)$.

2. Numerically integrate y^i forward with the assumption that $H_{uu}(y^i(t), t) > 0$.

3. At the terminal boundary, let

$$\lambda^i(t_f) - \phi_x(y^i_f) = \beta. \tag{5.133}$$

4. Linearize the first-order necessary conditions. For the control, this is

$$\begin{aligned} \Delta H_u(y^i(t), u^i, t) &= H_{ux}(y^i(t), u^i, t)\delta x(t) + H_{u\lambda}(y^i(t), u^i, t)\delta\lambda(t) \\ &\quad + H_{uu}(y^i(t), u^i, t)\delta u(t) + \mathcal{O}(\epsilon_1, \epsilon_2, \epsilon_3), \end{aligned} \tag{5.134}$$

so that

$$\begin{aligned} \delta H_u(y^i(t), u^i(t), t) &= H_{ux}(y^i(t), u^i(t), t)\delta x(t) + H_{u\lambda}(y^i(t), u^i(t), t)\delta\lambda(t) \\ &\quad + H_{uu}(y^i(t), u^i, t)\delta u(t) = 0. \end{aligned} \tag{5.135}$$

The dynamic equation is linearized as

$$\delta\dot{x}(t) = f_x(x^i(t), u^i(t), t)\delta x + f_u(x^i(t), u^i(t), t)\delta u(t). \tag{5.136}$$

For the Lagrange multipliers,

$$-\delta\dot{\lambda}(t) = H_{xx}(y^i(t), u^i(t), t)\delta x + H_{xu}(y^i(t), u^i(t), t)\delta u(t) + H_{x\lambda}(y^i(t), u^i(t), t)\delta\lambda(t). \tag{5.137}$$

From (5.135), solve for the change in control as

$$\delta u(t) = -H_{uu}^{-1}(y^i(t), u^i(t), t) \left[H_{ux}(y^i(t), u^i(t), t)\delta x(t) + H_{u\lambda}(y^i(t), u^i(t), t)\delta\lambda(t) \right]. \tag{5.138}$$

(Recall that $H_{uu}(y^i(t), u^i(t), t) > 0$ is assumed, so the inverse exists.) Combining (5.136), (5.137), and (5.138) into a matrix linear ordinary differential equation called the Hamiltonian system as given in (5.48):

$$\begin{bmatrix} \delta\dot{x} \\ \delta\dot{\lambda} \end{bmatrix} = \begin{bmatrix} A(t) - B(t)R^{-1}(t)C(t) & -B(t)R^{-1}(t)B^T(t) \\ -Q(t) + C^T(t)R^{-1}(t)C(t) & -(A(t) - B(t)R^{-1}(t)C(t))^T \end{bmatrix} \begin{bmatrix} \delta x \\ \delta\lambda \end{bmatrix}. \tag{5.139}$$

5. Solve numerically the Hamiltonian system from (5.139) as

$$\begin{bmatrix} \delta x(t_f) \\ \delta\lambda(t_f) \end{bmatrix} = \begin{bmatrix} \Phi_{11}(t_f, t_0) & \Phi_{12}(t_f, t_0) \\ \Phi_{21}(t_f, t_0) & \Phi_{22}(t_f, t_0) \end{bmatrix} \begin{bmatrix} \delta x(t_0) \\ \delta\lambda(t_0) \end{bmatrix}, \tag{5.140}$$

where, since $x(t_0)$ is given, $\delta x(t_0) = 0$. Use

$$\delta\lambda(t_f) - \phi_{xx}(t_f, t_0)\delta x(t_f) = d\beta \tag{5.141}$$

for $|\beta| > |\beta + d\beta|$ so that $d\beta$ is chosen such that β contracts on each iteration. Solving for the partitioned elements of (5.140)

$$\delta x(t_f) = \Phi_{12}(t_f, t_0)\delta\lambda(t_0), \tag{5.142}$$

$$\delta\lambda(t_f) = \Phi_{22}(t_f, t_0)\delta\lambda(t_0). \tag{5.143}$$

Substitute (5.142) and (5.143) into (5.141) to give

$$(\Phi_{22}(t_f, t_0) - \phi_{xx}(t_f, t_0)\Phi_{12}(t_f, t_0))\delta\lambda(t_0) = d\beta, \tag{5.144}$$

$$\Rightarrow \quad \delta\lambda(t_0) = [\Phi_{22}(t_f, t_0) - \phi_{xx}(t_f, t_0)\Phi_{12}(t_f, t_0)]^{-1} d\beta. \tag{5.145}$$

6. Update the guess of the initial Lagrange multipliers

$$\lambda^{i+1}(t_0) = \lambda^i(t_0) + \delta\lambda(t_0). \tag{5.146}$$

7. If $\beta < \varepsilon$ for some ε *small enough*, then stop. If not, go to 2.

Sophisticated numerical optimization algorithms based on the shooting method can be found in [43].

An Application of the Shooting Method

We apply the shooting method to the Brachistochrone problem of Section 3.3.3. Since an analytic solution has been obtained, the convergence rate of the numerical scheme can be accurately assessed. For this problem,

$$f_x = \begin{bmatrix} 0 & \cos\theta \\ 0 & 0 \end{bmatrix}, \qquad f_u = \begin{bmatrix} -v\sin\theta \\ g\cos\theta \end{bmatrix},$$

$$H_x = \begin{bmatrix} 0 & \lambda_r \cos\theta \end{bmatrix}, \qquad H_{xu} = \begin{bmatrix} 0 \\ -\lambda_r \sin\theta \end{bmatrix} = H_{ux}^T, \qquad H_{xx} = \begin{bmatrix} 0 & 0 \\ 0 & 0 \end{bmatrix},$$

$$H_{uu} = -\lambda_r v\cos\theta - \lambda_v g \sin\theta.$$

The terminal penalty function is

$$\phi(x(t_f)) = -r(t_f),$$

giving

$$\phi_x = \begin{bmatrix} -1 & 0 \end{bmatrix} \qquad \text{and} \qquad \phi_{xx} = \begin{bmatrix} 0 & 0 \\ 0 & 0 \end{bmatrix}.$$

Note that using (3.68), we could remove the differential equation for λ_r. It is left in here for simplicity.

An initial guess of $\lambda_r^0(0) = -1$ and $\lambda_v^0(0) = -0.8$ provides the results

$$r(t_f) = 9.350 \qquad \text{and} \qquad \lambda_v(t_f) = 0.641.$$

The transition matrix of (5.140) is computed for $t_f = 2$ and take $g = 9.80665$ to be

$$\Phi(2,0) = \begin{bmatrix} 1.0000 & 0.2865 & 13.0710 & -16.3410 \\ 0.0000 & -0.8011 & 20.4083 & -25.5103 \\ 0.0000 & 0.0000 & 1.0000 & 0.0000 \\ 0.0000 & 0.0610 & -1.1969 & 0.6950 \end{bmatrix}.$$

Using $d\beta = -\beta$, the desired change in the initial values of the Lagrange multipliers is computed as

$$\delta\lambda_0 = \begin{bmatrix} 0. \\ -0.9221309 \end{bmatrix}.$$

Note that as we already knew the correct value for λ_r, the desired update is zero.

Three more iterations provide

	$\lambda_v(0)$	$\lambda_v(t_f)$	$r(t_f)$
0	-0.8	0.6409	9.3499
1	-1.7221	-0.6856	11.5777
2	-1.2537	0.0307	12.4832
3	-1.2732	-0.0000	12.4862

On each step, the computed update for $\lambda_r(0)$ is zero, reflecting its invariance. Moreover, as expected, $\lambda_v(0)$ and $\lambda_v(2)$ get closer to their analytical initial and terminal values of $-\frac{4}{\pi}$ and 0, respectively.

5.5 LQ Problem with Linear Terminal Constraints: Transition Matrix Approach

The first-order necessary conditions of Theorem 5.3.1 are now used for the optimal control problem of (5.17) to (5.20), where the terminal constraints (5.20) and the boundary condition (5.43) are explicitly included. The two-point boundary-value problem of Theorem 5.3.1 is solved in a manner similar to that of Section 5.4 by using the transition matrix (5.50) as

$$x^o(t_f) = \Phi_{11}(t_f, t_0)x(t_0) + \Phi_{12}(t_f, t_0)\lambda(t_0), \tag{5.147}$$

$$\lambda(t_f) = \Phi_{21}(t_f, t_0)x(t_0) + \Phi_{22}(t_f, t_0)\lambda(t_0) = D^T\nu + S_f x^o(t_f). \tag{5.148}$$

From (5.148) the relation between $\lambda(t_0)$ and $(\nu, x(t_0))$, assuming that $\bar{\Phi}_{22}(t_f, t_0)$ defined in (5.72) is invertible, is

$$\lambda(t_0) = -\bar{\Phi}_{22}^{-1}(t_f, t_0)\bar{\Phi}_{21}(t_f, t_0)x(t_0) + \bar{\Phi}_{22}^{-1}(t_f, t_0)D^T\nu. \tag{5.149}$$

This is introduced into (5.147), which in turn is introduced into (5.20), producing

$$\begin{aligned} Dx^o(t_f) &= D\left[\bar{\Phi}_{11}(t_f,t_0) - \bar{\Phi}_{12}(t_f,t_0)\bar{\Phi}_{22}^{-1}(t_f,t_0)\bar{\Phi}_{21}(t_f,t_0)\right]x(t_0) \\ &\quad + D\bar{\Phi}_{12}(t_f,t_0)\bar{\Phi}_{22}^{-1}(t_f,t_0)D^T\nu = 0. \end{aligned} \tag{5.150}$$

The symplectic property of the Hamiltonian transition matrix is used to reduce the coefficient of x_0 in (5.150). By using the symplectic identity (5.71) for the symplectic matrix $\bar{\Phi}_H(t_f,t_0)$ defined in (5.72) and the assumed invertibility of $\bar{\Phi}_{22}(t_f,t_0)$, we obtain

$$\bar{\Phi}_{11}(t_f,t_0) - \bar{\Phi}_{12}(t_f,t_0)\bar{\Phi}_{21}^T(t_f,t_0)\bar{\Phi}_{22}^{-T}(t_f,t_0) = \bar{\Phi}_{22}^{-T}(t_f,t_0). \tag{5.151}$$

By premultiplying and postmultiplying the symplectic identity (5.70) by $\bar{\Phi}_{22}^{-1}(t_f,t_0)$ and $\bar{\Phi}_{22}^{-T}(t_f,t_0)$, respectively, we obtain

$$\bar{\Phi}_{22}^{-1}(t_f,t_0)\bar{\Phi}_{21}(t_f,t_0) = \bar{\Phi}_{21}^T(t_f,t_0)\bar{\Phi}_{22}^{-T}(t_f,t_0). \tag{5.152}$$

Using (5.152) in (5.151), the coefficient of x_0 in (5.150) reduces to

$$\bar{\Phi}_{11}(t_f,t_0) - \bar{\Phi}_{12}(t_f,t_0)\bar{\Phi}_{22}^{-1}(t_f,t_0)\bar{\Phi}_{21}(t_f,t_0) = \bar{\Phi}_{22}^{-T}(t_f,t_0). \tag{5.153}$$

Finally, by solving the matrix equation (5.153) for $\bar{\Phi}_{11}(t_f,t_0)$, then eliminating $\bar{\Phi}_{11}(t_f,t_0)$ in (5.69), and using the symmetric property established in (5.152), the symmetric property of $\bar{\Phi}_{12}(t_f,t_0)\bar{\Phi}_{22}^{-1}(t_f,t_0)$ is also established. Therefore, (5.149) and (5.150) can be written in the symmetric form

$$\begin{bmatrix} \lambda(t_0) \\ 0 \end{bmatrix} = \begin{bmatrix} -\bar{\Phi}_{22}^{-1}(t_f,t_0)\bar{\Phi}_{21}(t_f,t_0) & \bar{\Phi}_{22}^{-1}(t_f,t_0)D^T \\ D\bar{\Phi}_{22}^{-T}(t_f,t_0) & D\bar{\Phi}_{12}(t_f,t_0)\bar{\Phi}_{22}^{-1}(t_f,t_0)D^T \end{bmatrix} \begin{bmatrix} x_0 \\ \nu \end{bmatrix}. \tag{5.154}$$

At this point, $\lambda(t_0)$ and ν can be determined as a function of $x(t_0)$. Substitution of this result into (5.46) will produce an optimal control rule for the terminal constrained optimal control problem not only for t_0 but for all $t \in [t_0, t_f]$. Before explicitly doing this, the elements of the coefficient matrix of (5.154), identified as

$$S(t_f, t; S_f) \triangleq -\bar{\Phi}_{22}^{-1}(t_f, t)\bar{\Phi}_{21}(t_f, t), \tag{5.155}$$

$$F^T(t_f, t) \triangleq \bar{\Phi}_{22}^{-1}(t_f, t)D^T, \tag{5.156}$$

$$G(t_f, t) \triangleq D\bar{\Phi}_{12}(t_f, t)\bar{\Phi}_{22}^{-1}(t_f, t)D^T, \tag{5.157}$$

are to be analyzed. Our objective is to find the propagation equations for these elements and discuss their properties. In particular, these elements will be seen to combine into a Riccati variable for the constrained control problem, and the coefficient matrix of (5.154) is a symmetric matrix.

As given in Theorem 5.4.1, $S(t_f, t; S_f)$ satisfies the Riccati differential equation (5.78). The differential equation for $F(t_f, t)$ is now developed by determining the differential equation for $\bar{\Phi}_{22}^{-1}(t_f, t)$ by noting

$$\frac{d}{dt}\left[\bar{\Phi}_{22}^{-1}(t_f, t)\right]\bar{\Phi}_{22}(t_f, t) + \bar{\Phi}_{22}^{-1}(t_f, t)\frac{d}{dt}\left[\bar{\Phi}_{22}(t_f, t)\right] = 0. \tag{5.158}$$

From the adjoint differential equation for $\Phi_H(t_f, t)$ (5.76), (5.158) becomes

$$\begin{aligned}\frac{d}{dt}\left[\bar{\Phi}_{22}^{-1}(t_f, t)\right]\bar{\Phi}_{22}(t_f, t) &= -\bar{\Phi}_{22}^{-1}(t_f, t)\Big[\bar{\Phi}_{21}(t_f, t)B(t)R(t)^{-1}B^T(t) \\ &\quad + \bar{\Phi}_{22}(t_f, t)\left(A(t) - B(t)R(t)^{-1}C(t)\right)^T\Big]. \end{aligned} \tag{5.159}$$

Then, by the assumed invertibility of $\bar{\Phi}_{22}(t_f, t)$, using (5.155)

$$\begin{aligned}&\frac{d}{dt}\left[\bar{\Phi}_{22}^{-1}(t_f, t)\right] \\ &= -\Big[\left(A(t) - B(t)R^{-1}(t)C(t)\right)^T - S(t_f, t; S_f)B(t)R^{-1}(t)B^T(t)\Big]\bar{\Phi}_{22}^{-1}(t_f, t).\end{aligned} \tag{5.160}$$

Noting that $\frac{d}{dt}\left[F^T(t_f, t)\right] = \frac{d}{dt}\left[\bar{\Phi}_{22}^{-1}(t_f, t)\right]D^T$, then

$$\begin{aligned}&\frac{d}{dt}\left[F^T(t_f, t)\right] \\ &= -\Big[\left(A(t) - B(t)R^{-1}(t)C(t)\right)^T - S(t_f, t; S_f)B(t)R^{-1}(t)B^T(t)\Big]F^T(t_f, t), \\ &\quad F^T(t_f, t_f) = D^T.\end{aligned} \tag{5.161}$$

In a similar manner, the differential equation for $G(t_f, t)$ is obtained by direct differentiation of $\bar{\Phi}_{12}(t_f, t)\bar{\Phi}_{22}^{-1}(t_f, t)$, as

$$\begin{aligned}
&\frac{d}{dt}\left[\bar{\Phi}_{12}(t_f, t)\right]\bar{\Phi}_{22}^{-1}(t_f, t) + \bar{\Phi}_{12}(t_f, t)\frac{d}{dt}\left[\bar{\Phi}_{22}^{-1}(t_f, t)\right] \\
&= \Big\{\Big[\bar{\Phi}_{11}(t_f, t)B(t)R^{-1}(t)B(t) + \bar{\Phi}_{12}(t_f, t)\left(A(t) - B(t)R^{-1}(t)C(t)\right)^T\Big] \\
&\quad - \bar{\Phi}_{12}(t_f, t)\Big[\left(A(t) - B(t)R^{-1}(t)C(t)\right)^T - S(t_f, t; S_f)B(t)R^{-1}(t)B^T(t)\Big]\Big\} \\
&\qquad \times \bar{\Phi}_{22}(t_f, t)^{-1} \\
&= \left[\bar{\Phi}_{11}(t_f, t) - \bar{\Phi}_{12}(t_f, t)\bar{\Phi}_{22}^{-1}(t_f, t)\bar{\Phi}_{21}(t_f, t)\right]B(t)R^{-1}(t)B^T(t)\bar{\Phi}_{22}^{-1}(t_f, t). \qquad (5.162)
\end{aligned}$$

Equation (5.162) is reduced further, by using the symplectic identity of (5.153), to

$$\frac{d}{dt}\left[\bar{\Phi}_{12}(t_f, t)\bar{\Phi}_{22}^{-1}(t_f, t)\right] = \bar{\Phi}_{22}^{-T}(t_f, t)B(t)R(t)^{-1}B^T(t)\bar{\Phi}_{22}^{-1}(t_f, t). \qquad (5.163)$$

By premultiplying and postmultiplying (5.163) by D and D^T, respectively, and using the definitions of $F(t_f, t)$ and $G(t_f, t)$, the differential equation for $G(t_f, t)$ is

$$\dot{G}(t_f, t) = F(t_f, t)B(t)R^{-1}(t)B^T(t)F^T(t_f, t), \quad G(t_f, t_f) = 0. \qquad (5.164)$$

Note that $G(t_f, t)$ generated by (5.164) is symmetric. Since it was already shown that $S(t_f, t; S_f)$ is symmetric, the coefficient matrix of (5.154) is symmetric.

Our objective is to determine ν in terms of $x(t_0)$. This can occur only if $G(t_f, t_0)$ is invertible. For this to happen it is necessary for D to be full rank. Assuming $G(t_f, t_0)$ is invertible,

$$\nu = -G^{-1}(t_f, t_0)F(t_f, t_0)x(t_0). \qquad (5.165)$$

The invertibility of $G(t_f, t_0)$ is known as a normality condition, ensuring a finite ν for a finite $x(t_0)$. By using (5.165) to eliminate ν in (5.149),

$$\lambda(t_0) = \bar{S}(t_f, t_0)x(t_0), \qquad (5.166)$$

where

$$\bar{S}(t_f, t) = S(t_f, t; S_f) - F^T(t_f, t)G^{-1}(t_f, t)F(t_f, t). \tag{5.167}$$

The optimal control can now be written as an explicit function of the initial state. If t_0 is considered to be the present time t, then introducing (5.166) and (5.167) into (5.46) results in the optimal control rule for the terminal constrained optimal control problem as

$$u^o(t) = -R^{-1}(t)[C(t) + B^T(t)\bar{S}(t_f, t)]x(t). \tag{5.168}$$

$\bar{S}(t_f, t)$ satisfies the same Riccati differential equation as (5.78). This can be verified by time differentiation of $\bar{S}(t_f, t)$ defined in (5.167). Furthermore, if all the terminal states are constrained, i.e., $D = I$, and (5.155) to (5.157) are used in (5.167), then by the symplectic identities, $\bar{S}(t_f, t)$ reduces to

$$\bar{S}(t_f, t) = -\bar{\Phi}_{12}^{-1}(t_f, t)\bar{\Phi}_{11}(t_f, t). \tag{5.169}$$

The major difficulty with propagating $\bar{S}(t_f, t)$ directly is in applying the proper boundary conditions at t_f. From (5.167), $\bar{S}(t_f, t_f)$ is not defined because $G(t_f, t_f)$ is not invertible. The integration of $S(t_f, t; S_f)$, $F(t_f, t)$, and $G(t_f, t)$ may have a computational savings over the integration of the transition matrix. Furthermore, $G(t_f, t)$ and $F(t_f, t)$ do not have to be integrated over the entire interval $[t_0, t_f]$ but only until $G(t_f, t)$ is invertible, usually some very small time step away from t_f. This allows a proper initialization for $\bar{S}(t_f, t)$. Once $\bar{S}(t_f, t)$ is formed, only $\bar{S}(t_f, t)$ need be propagated backward in time.

The behavior of $\bar{S}(t_f, t)$ is reflected in the behavior of $u^o(t)$ near t_f. For large deviations away from the terminal manifold, $u^o(t)$ reacts by emphasizing the satisfaction of the constraints rather than reducing the performance criterion.

In the next subsection, we demonstrate that the invertibility of $G(t_f, t)$ is equivalent to a controllability requirement associated only with the required terminal boundary restriction (5.20).

5.5.1 Normality and Controllability for the LQ Problem

We show here that the normality condition assumed in (5.165) is actually a controllability requirement. This is done by converting the original problem to one in which the quadratic cost criterion is only a function of a control variable. The minimization of this new performance criterion subject to given initial conditions and the terminal constraints (5.20) requires a controllability condition, which is just $G(t_f, t)$. The following theorem is similar to that of Brockett [8] and others.

Theorem 5.5.1 *Assume that the symmetric matrix $S(t_f, t; S_f)$, which is a solution to the Riccati equation (5.78), exists on the interval $t_0 \le t \le t_f$. Then there exists a control $u(\cdot)$ on the interval $t_0 \le t \le t_f$ that minimizes (5.17) subject to the differential constraint (5.18) and the boundary conditions (5.19) and (5.20) if and only if there exists a $v(\cdot)$ on the interval $t_0 \le t \le t_f$ which minimizes*

$$J_1(v(\cdot); x_0, t) = \frac{1}{2}\int_{t_0}^{t_f} v^T(t)R(t)v(t)dt \tag{5.170}$$

subject to the differential constraint

$$\dot{x} = \underline{A}(t)x(t) + B(t)v(t), \tag{5.171}$$

where

$$\underline{A}(t) \triangleq A(t) - B(t)R^{-1}(t)\left(C(t) + B^T(t)S(t_f, t; S_f)\right) \tag{5.172}$$

with the boundary conditions $x(t_0) = x_0$ and $Dx(t_f) = 0$.

Proof: By proceeding exactly as was done to obtain (5.96), we note that if we let

$$v(t) = u(t) + R^{-1}(t)\left(C(t) + B^T(t)S(t_f, t; S_f)\right)x(t), \tag{5.173}$$

then the cost specified in (5.96) can be written as

$$J\left(v(\cdot);x_0,t\right)=\frac{1}{2}x^T(t_0)S(t_f,t_0;S_f)x(t_0)+\frac{1}{2}\int_{t_0}^{t_f}v^T(t)R(t)v(t)dt. \quad (5.174)$$

Since $x(t_0)=x_0$ is given, the cost function upon which $v(\cdot)$ has influence is

$$J_1\left(v(\cdot);x_0,t\right)=\frac{1}{2}\int_{t_0}^{t_f}v^T(t)R(t)v(t)dt, \quad (5.175)$$

which is subject to the differential constraint (5.171) when (5.173) is substituted into (5.18). ▮

We now proceed to solve this accessory problem of minimizing (5.170) subject to (5.171) using the technique given at the beginning of this section. First, the Riccati variable $S^*(t_f,t;0)$ is propagated by (5.78) as

$$\begin{aligned}\dot{S}^*(t_f,t;0) &= -\underline{A}(t)S^*(t_f,t;0)-S^*(t_f,t;0)\underline{A}^T(t)\\ &\quad +S^*(t_f,t;0)B(t)R^{-1}(t)B^T(t)S^*(t_f,t;0), \quad (5.176)\\ S^*(t_f,t_f;0) &= 0,\end{aligned}$$

where the $*$ superscript is used to denote dependence on the Riccati variable $S(t_f,t;S_f)$. For this problem, $Q(t)$ and $C(t)$ are now zero. The solution to this homogeneous Riccati equation with zero initial condition is $S^*(t_f,t;0)=0$ over the interval $t_0\le t\le t_f$. The propagation of the linear differential equation (5.161) is

$$F^T(t_f,t)=\Phi_{\underline{A}}^T(t_f,t)D^T. \quad (5.177)$$

Using (5.177) in (5.164), the solution to $G(t_f,t)$ is

$$G(t_f,t_0)=-DW_{\underline{A}}(t_f,t_0)D^T, \quad (5.178)$$

where

$$W_{\underline{A}}(t_f,t_0)=\int_{t_0}^{t_f}\Phi_{\underline{A}}(t_f,t)B(t)R^{-1}(t)B^T(t)\Phi_{\underline{A}}^T(t_f,t)dt \quad (5.179)$$

is the controllability Grammian. Therefore, the invertibility of $G(t_f, t_0)$ does not depend upon controllability of the entire state space but depends only on the controllability to the desired terminal manifold $Dx(t_f) = 0$. Clearly, the invertibility of $G(t_f, t_0)$ is a controllability condition for $v(t)$ to reach the terminal manifold. Then, by (5.173), invertibility of $G(t_f, t_0)$ is also a controllability condition on $u(t)$. Note that $F(t_f, t)$ and $G(t_f, t_0)$ are the same as those that would be generated by the original problem.

Theorem 5.5.2 *If*

$$\dot{x}(t) = A(t)x(t) + B(t)u(t) \tag{5.180}$$

is controllable on the interval $[t_0, t_f]$ *and if*

$$u(t) = v(t) + \Lambda(t)x(t), \tag{5.181}$$

then

$$\dot{x}(t) = (A(t) + B(t)\Lambda(t))x(t) + B(t)v(t) = \underline{A}(t)x(t) + B(t)v(t) \tag{5.182}$$

is also controllable for any finite piecewise continuous $\Lambda(t)$ *on the interval* $[t_0, t_f]$.

Proof: We need to show that $x_0^T \Phi_{\underline{A}}(t_0, t)B(t) = 0$ for all t in the interval implies that $x_0^T \Phi_A(t_0, t)B(t) = 0$, a contradiction to the controllability assumption. To do this, note that the controllability Grammian

$$W_A(t_0, t_f) = \int_{t_0}^{t_f} \Phi_A(t_0, t)B(t)B^T(t)\Phi_A^T(t_0, t)dt \tag{5.183}$$

can be obtained by integrating the linear matrix equation

$$\begin{aligned} \dot{W}_A(t, t_f) &= A(t)W_A(t, t_f) + W_A(t, t_f)A^T(t) - B(t)B^T(t), \qquad (5.184) \\ W_A(t_f, t_f) &= 0. \end{aligned}$$

Similarly, the controllability Grammian (5.179) for $\underline{A}(t)$ is obtained from

$$\begin{aligned} \dot{W}_{\underline{A}}(t,t_f) &= \underline{A}(t)W_{\underline{A}}(t,t_f) + W_{\underline{A}}(t,t_f)\underline{A}^T(t) - B(t)B^T(t), \quad (5.185) \\ W_{\underline{A}}(t_f,t_f) &= 0. \end{aligned}$$

Form the matrix

$$E(t) = W_{\underline{A}}(t,t_f) - W_A(t,t_f). \tag{5.186}$$

It has a linear differential equation

$$\begin{aligned} \dot{E}(t) &= \underline{A}(t)E(t) + E(t)\underline{A}^T(t) + B(t)\Lambda(t)W_A(t,t_f) \\ &\quad + W_A(t,t_f)\Lambda^T(t)B^T(t), \; E(t_f) = 0. \end{aligned} \tag{5.187}$$

Then

$$\begin{aligned} E(t_0) &= -\int_{t_0}^{t_f} \Phi_{\underline{A}}(t_0,t)\Big[B(t)\Lambda(t)W_A(t,t_f) \\ &\quad + W_A(t,t_f)\Lambda^T(t)B^T(t)\Big]\Phi_{\underline{A}}^T(t_0,t)dt. \end{aligned} \tag{5.188}$$

Since by hypothesis, $x_0^T\Phi_{\underline{A}}(t_0,t)B(t) = 0$, then $x_0^T E(t_0)x_0 = 0$ by (5.188) and $x_0^T W_{\underline{A}}(t_0,t)x_0 = 0$ by (5.179). But from (5.186) evaluated at t_0 this implies that

$$x_0^T W_A(t_0,t_f)x_0 = 0, \tag{5.189}$$

which is a contradiction of the assumed controllability of

$$\dot{x}(t) = A(t)x(t) + B(t)u(t). \tag{5.190}$$

∎

From Theorem 5.5.2 we see that the controllability Assumption 5.3.1 implies the normality condition $G(t_f,t_0) < 0$. This observation implies that state feedback as given in (5.181) does not change controllability. When the controllability Assumption 5.3.1 for reaching the manifold (5.20) is restricted to complete controllability, then

$$G(t_f,t) < 0 \quad \forall t \in [t_0,t_f). \tag{5.191}$$

5.5.2 Necessary and Sufficient Conditions for the Positivity of the Terminally Constrained Quadratic Cost Criterion

The objective of this section is to show that a necessary and sufficient condition for the quadratic cost criterion (5.17) with linear terminal constraints (5.20) to be positive definite when $x(t_0) = 0$ is that the Riccati variable $\bar{S}(t_f, t)$ exist for all t in the interval $[t_0, t_f)$. The results of this section closely parallel those of Section 5.4.4 for the unconstrained problem.

First, it is shown that if there exists an $\bar{S}(t_f, t)$ defined by (5.167) for all t in the interval $[t_0, t_f)$, then $J(u(\cdot); 0, t_o)$ is positive definite. Consider adding to $J(u(\cdot); 0, t_0)$ of (5.17) the identically zero quantity

$$-\frac{1}{2}\begin{bmatrix} x^T(t), & \nu^T \end{bmatrix}\begin{bmatrix} S(t_f, t; S_f) & F^T(t_f, t) \\ F(t_f, t) & G(t_f, t) \end{bmatrix}\begin{bmatrix} x(t) \\ \nu \end{bmatrix}\Bigg|_{t_0}^{t_f} \tag{5.192}$$
$$+\frac{1}{2}\int_{t_0}^{t_f} \frac{d}{dt}\left\{\begin{bmatrix} x^T(t), & \nu^T \end{bmatrix}\begin{bmatrix} S(t_f, t; S_f) & F^T(t_f, t) \\ F(t_f, t) & G(t_f, t) \end{bmatrix}\begin{bmatrix} x(t) \\ \nu \end{bmatrix}\right\} dt = 0.$$

By using the Riccati equation (5.78), the propagation equation for $F(t_f, t)$ in (5.161), and that of $G(t_f, t)$ in (5.164), the cost criterion (5.17) can be manipulated into a perfect square as

$$\begin{aligned} J(u(\cdot); x(t_0), t_0) &= \frac{1}{2}x^T(t_0)S(t_f, t_0; S_f)x(t_0) \\ &\quad + x^T(t_0)F^T(t_f, t_0)\nu + \frac{1}{2}\nu^T G(t_f, t_0)\nu - x^T(t_f)D^T\nu \\ &\quad + \int_{t_0}^{t_f} \big\| R^{-1}(t)\left[\left(C(t) + B^T(t)S(t_f, t; S_f)\right)x(t)\right. \\ &\quad + \left. B^T(t)F^T(t_f, t)\nu\right] + u(t)\big\|_{R(t)}^2 \, dt. \end{aligned} \tag{5.193}$$

One difficulty occurs at $t = t_f$, where ν cannot be determined in terms of $x(t_f)$. Since the system is assumed completely controllable by Assumption 5.4.1, then $G(t_f, t_f - \Delta)$ is invertible for any $\Delta > 0$. For small enough Δ, $S(t_f, t_f - \Delta; S_f)$

exists. Therefore, the optimal control in an interval $t_f - \Delta \leq t \leq t_f$ is open-loop over that interval, given by

$$
\begin{aligned}
u^o(t) \; = \; & -R^{-1}(t) \Big\{ \left[C(t), \;\; B^T(t) \right] \Phi_H(t, \;\; t_f - \Delta) \begin{bmatrix} I \\ S(t_f, t_f - \Delta; S_f) \end{bmatrix} \\
& - \; B^T(t) F^T(t_f, t) G^{-1}(t_f, t_f - \Delta) F(t_f, t_f - \Delta) \Big\} x(t_f - \Delta), \qquad (5.194)
\end{aligned}
$$

where ν is determined by (5.165) evaluated at $t_0 = t_f - \Delta$. Given Assumption 5.4.1, this open-loop control will satisfy the terminal boundary conditions. Since all the factors in (5.194) remain finite in $[t_f - \Delta, t_f]$, $u^o(t)$ remains finite in that interval. In the interval $t_0 \leq t \leq t_f - \Delta$ the optimal control is given by (5.168). By using the optimal control given by (5.194) for $t_f - \Delta \leq t \leq t_f$ and (5.168) for $t_0 \leq t \leq t_f - \Delta$ to replace the control $u(\cdot)$, the integral part of the cost in (5.193) becomes zero and is at its minimum value. With $x(t_0) = 0$, the optimal control is the null control. Any other control which satisfies the terminal boundary condition that $Dx(t_f) = 0$ and is unequal to $u^o(\cdot)$ will give a positive value to $J\left(u(\cdot); 0, t_0\right)$.

The arguments for the existence of $\bar{S}(t_f, t)$ over the interval $[t_0, t_f)$, given that the cost $J\left(u(\cdot); 0, t_0\right)$ is positive definite, are the same as that given in Section 5.4.4. That is, for some t' close enough to t_f, $\bar{S}(t_f, t)$ exists for $t_0 \leq t < t_f$. Therefore, the optimal control laws (5.194) or (5.168) apply, and the optimal cost starting at some $x(t')$ is

$$
J\left(u^o(\cdot); x(t'), t'\right) = \frac{1}{2} x^T(t') \bar{S}(t_f, t') x(t'). \qquad (5.195)
$$

By applying a control suggested by (5.98), controllability, and positivity of the cost for $x(t_0) = 0$, the arguments of Section 5.4.4 imply that since $x^T(t')\bar{S}(t_f, t')x(t')$ can go to neither positive nor negative infinity for all finite $x(t')$ and for any t' in the interval $[t_0, t_f)$, then $\bar{S}(t_f, t')$ exists for all t' in $[t_0, t_f)$. These results are summarized as Theorem 5.5.3.

Theorem 5.5.3 *Given Assumption* 5.4.1, *a necessary and sufficient condition for* $J(u(\cdot); 0, t_0)$ *to be positive definite for the class of controls which satisfy the terminal constraint* (5.20) *is that there exist a function* $\bar{S}(t_f, t)$ *for all* t *in* $[t_0, t_f)$ *which satisfies the Riccati equation* (5.78).

Remark 5.5.1 *Construction of* $\bar{S}(t_f, t)$ *by* (5.167) *using its component parts may not be possible since* $S(t_f, t; S_f)$ *may not exist, whereas* $\bar{S}(t_f, t)$ *can exist.*

Remark 5.5.2 *Theorem* 5.5.3 *can be extended to show that* $J(u(\cdot); 0, t_0)$ *is strongly positive definite by a proof similar to that used in Theorem* 5.4.3 *of Section* 5.4.5.

Remark 5.5.3 *Note that the optimal value function for the LQ problem with terminal constraints is given as*

$$V(x(t), t) = \frac{1}{2} x(t)^T \bar{S}(t_f, t) x(t). \tag{5.196}$$

This satisfies the Hamilton–Jacobi–Bellman equation given in Theorem 4.7.1 *of Section* 4.7.

Example 5.5.1 (Shortest distance between two points on a sphere) *In this section the minimum distance problem is generalized from the unconstrained terminal problem, given in Example* 5.4.1, *to the terminal constrained problem of reaching a given terminal point. The problem is to find the control* u *that minimizes*

$$J = \int_0^{\phi_1} (u^2 + \cos^2\theta)^{\frac{1}{2}} d\phi \tag{5.197}$$

subject to (5.111) *at fixed terminal* $\phi = \phi_1$ *with constraint* $\theta(\phi_1) = 0$.

The first-order necessary conditions are

$$\frac{d\theta}{d\phi} = H_\lambda = u,\ \theta(0) = 0,\ \theta(\phi_1) = 0, \tag{5.198}$$

$$-\dot{\lambda} = H_\theta = (u^2 + \cos^2\theta)^{-\frac{1}{2}} \sin\theta\cos\theta,\ \lambda(\phi_1) = \nu, \tag{5.199}$$

$$H_u = \lambda + u(u^2 + \cos^2\theta)^{-\frac{1}{2}} = 0. \tag{5.200}$$

The first-order necessary conditions are satisfied by the trajectory $u(\phi) = 0$, $\lambda(\phi) = 0 (\nu = 0)$, $\theta(\phi) = 0$ *for* $0 \leq \phi \leq \phi_1$.

About this extremal path, the second-order necessary conditions are generated. The second variational problem given in Section 5.2 *is to find the perturbed control* δu *that minimizes*

$$2\delta^2 J = \int_0^{\phi_1} \left((\delta u)^2 - (\delta\theta)^2 \right) d\phi \tag{5.201}$$

subject to $\frac{d\delta\theta}{d\phi} = \delta u$, $\delta\theta(0) = 0$ *for fixed terminal* $\phi = \phi_1$ *with the terminal constraint* $\delta\theta(\phi_1) = 0$. *From Theorem* 5.5.3 *it is necessary that the solution of the associated Riccati equation exist. Since for this example* $A = 0$, $B = 1$, $C = 0$, $Q = -1$, $R = 1$, *the solution of the associated Riccati equation is obtained from the following equations and solutions:*

$$\frac{dS}{d\phi} = S^2 + 1,\ S(\phi_1) = 0 \Rightarrow S(\phi) = -\tan(\phi_1 - \phi), \tag{5.202}$$

$$\frac{dF}{d\phi} = SF,\ F(\phi_1) = 1 \Rightarrow F(\phi) = \sec(\phi_1 - \phi), \tag{5.203}$$

$$\frac{dG}{d\phi} = F^2,\ G(\phi_1) = 0 \Rightarrow G(\phi) = -\tan(\phi_1 - \phi). \tag{5.204}$$

From these solutions of $S(\phi)$, $F(\phi)$, *and* $G(\phi)$ *the associated Riccati solution is constructed as*

$$\bar{S}(\phi) = S(\phi) - F^2(\phi) G^{-1}(\phi) = \cot(\phi_1 - \phi). \tag{5.205}$$

Note that the solution remains finite until the solution escapes at $\phi_1 - \phi = \pi$. *At that point any great circle path from the point to the terminal point will give the same cost. If the path is longer than* π, *then there are paths that do not even satisfy the first-order necessary conditions that can give smaller cost. The second variational controller is*

$\delta u = -\cot(\phi_1 - \phi)\delta\theta$. *Note that the gain is initially positive, indicating that the best neighboring optimum controller is essentially divergent for* $\pi/2 < \phi_1 - \phi < \pi$ *but becomes convergent over the interval* $0 < \phi_1 - \phi \leq \pi/2$.

5.6 Solution of the Matrix Riccati Equation: Additional Properties

Three important properties of the solution to the matrix Riccati equation are given here. First, it is shown that if the terminal weight in the cost criterion is increased, the solution to the corresponding Riccati equation is also increased. This implies, for example, that $S(t_f, t; S_f) \leq \bar{S}(t_f, t)$. Second, by restricting certain matrices to be positive semidefinite, $S(t_f, t; S_f)$ is shown to be nonnegative definite and bounded. Finally, if $S_f = 0$ and the above restrictions hold, $S(t_f, t; 0)$ is monotonically increasing as t_f increases. This is extremely important for the next section, where we analyze the infinite-time, LQ problem with constant coefficients.

Theorem 5.6.1 *If* S_f^1 *and* S_f^2 *are two terminal weights in the cost criterion* (5.17)*, such that* $S_f^1 - S_f^2 \geq 0$*, then the difference* $S(t_f, t; S_f^1) - S(t_f, t; S_f^2) \geq 0$ *for all* $t \leq t_f$ *where* $S(t_f, t; S_f^2)$ *exists, and* $S(t_f, t; S_f^1)$ *and* $S(t_f, t; S_f^2)$ *are solutions of* (5.78) *for* S_f^1*,* S_f^2*, respectively.*

Proof: First, the unconstrained optimization problem of (5.17) and (5.18) is converted, as in Theorem 5.5.1, to an equivalent problem. This is done precisely in the manner used to obtain (5.96). However, now we are concerned with two problems having terminal weights S_f^1 and S_f^2, respectively. Therefore, (5.95) is indexed with a superscript 2 and added to a cost criterion having a terminal weight S_f^1. The result is that the optimal cost for weight S_f^1 is

$$\begin{aligned}
&\frac{1}{2}x^T(t_0)S(t_f,t_0;S_f^1)x(t_0)\\
&\quad = \frac{1}{2}x^T(t_0)S(t_f,t_0;S_f^2)x(t_0) + \min_{u(\cdot)}\left\{\frac{1}{2}x^T(t_f)(S_f^1 - S_f^2)x(t_f)\right. \qquad (5.206)\\
&\quad + \frac{1}{2}\int_{t_0}^{t_f}\left[u(t) + R^{-1}(t)\left(C(t) + B^T(t)S(t_f,t;S_f^2)\right)x(t)\right]^T\\
&\quad \left.\times R(t)\left[u(t) + R^{-1}(t)\left(C(t) + B^T(t)S(t_f,t;S_f^2)\right)x(t)\right]dt\right\}.
\end{aligned}$$

Clearly, a new problem results, if again

$$v(t) = u(t) + R^{-1}(t)\left(C(t) + B^T(t)S(t_f,t;S_f^2)\right)x(t), \tag{5.207}$$

where it is assumed that $S(t_f,t;S_f^2)$ exists, as

$$\begin{aligned}
&x_0^T\left[S(t_f,t_0;S_f^1) - S(t_f,t_0;S_f^2)\right]x_0\\
&\quad = \min_{v(\cdot)}\left[x^T(t_f)\left(S_f^1 - S_f^2\right)x(t_f) + \frac{1}{2}\int_{t_0}^{t_f} v^T(t)R(t)v(t)dt\right].
\end{aligned} \tag{5.208}$$

Since $S_f^1 - S_f^2$ is nonnegative definite, then the optimal cost must be nonnegative, implying that

$$x_0^T\left[S(t_f,t_0;S_f^1) - S(t_f,t_0;S_f^2)\right]x_0 \geq 0 \tag{5.209}$$

for all x_0 and $t_0 \leq t_f$. ▌

Remark 5.6.1 *Since this new control problem* (5.208) *will have an optimal cost as* $x_0^T S^*(t_f,t;S_f^1 - S_f^2)x_0$, *then a relationship exists as*

$$S(t_f,t_0;S_f^1) = S(t_f,t_0;S_f^2) + S^*(t_f,t_0;S_f^1 - S_f^2), \tag{5.210}$$

where $S^*(t_f,t;S_f^1 - S_f^2)$ *satisfies the homogeneous Riccati equation*

$$\begin{aligned}
\frac{d}{dt}S^*(t_f,t;S_f^1 - S_f^2) &= -\underline{A}(t)S^*(t_f,t;S_f^1 - S_f^2) - S^*(t_f,t;S_f^1 - S_f^2)\underline{A}^T(t)\\
&\quad S^*(t_f,t;S_f^1 - S_f^2)B(t)R^{-1}(t)B^T(t)S^*(t_f,t;S_f^1 - S_f^2),\\
&\quad S^*(t_f,t_f;S_f^1 - S_f^2) = S_f^1 - S_f^2,
\end{aligned} \tag{5.211}$$

with

$$\underline{A}(t) = A(t) - B(t)R^{-1}(t)\left(C(t) + B^T(t)S(t_f, t_0; S_f^2)\right). \tag{5.212}$$

Note that if $S^*(t_f, t; S_f^1 - S_f^2)$ has an inverse, then a linear matrix differential equation for $S^{*-1}(t_f, t; S_f^1 - S_f^2)$ results by simply differentiating $S^*(t_f, t; S_f^1 - S_f^2)S^{*-1}(t_f, t; S_f^1 - S_f^2) = I$.

Remark 5.6.2 *From (5.167), the difference between the constrained Riccati matrix $\bar{S}(t_f, t)$ and the unconstrained Riccati matrix $S(t_f, t; S_f)$ is*

$$\bar{S}^*(t_f, t) \triangleq -F^T(t_f, t)G^{-1}(t_f, t)F(t_f, t) \geq 0 \tag{5.213}$$

for $t \in [t_0, t_f)$, since $G(t_f, t) < 0$ for $t \in [to, t_f)$ by virtue of Theorem 5.5.2. Furthermore, the differential equation for $\bar{S}^*(t_f, t)$ is that of (5.211), but the boundary condition at t_f is not defined. Intuitively, the constrained problem can be thought of as a limit of the unconstrained problem, where certain elements of the weighting function are allowed to go to infinity.

Existence of the solution to the Riccati equation is of central importance. Results using the following restrictive but useful assumption guarantees not only that $S(t_f, t; S_f)$ exists but that it is nonnegative definite.

Assumption 5.6.1 *$S_f \geq 0$ and $Q(t) - C^T(t)R^{-1}(t)C(t) \geq 0$ for all t in the interval $[t_0, t_f]$.*

Theorem 5.6.2 *Given Assumptions 5.2.1, 5.4.1, and 5.6.1, the solution to the Riccati equation (5.78), $S(t_f, t; S_f)$, exists on the interval $t_0 \leq t \leq t_f$, regardless of t_0, and is nonnegative definite.*

Proof: From (5.97), the minimum cost is related to the Riccati variable, if it exists, for an arbitrary initial state as

$$\frac{1}{2}x^T(t_0)S(t_f,t_0;S_f)x(t_0) = \min_{u(\cdot)}\left\{\frac{1}{2}x^T(t_f)S_f x(t_f)\right. \tag{5.214}$$

$$\left. + \frac{1}{2}\int_{t_0}^{t_f}\left[x^T(t)Q(t)x(t) + 2u^T(t)C(t)x(t) + u^T(t)R(t)u(t)\right]dt\right\}.$$

Let us make a change in controls of the form

$$u(t) = v(t) - R^{-1}(t)C(t)x(t). \tag{5.215}$$

The cost can now be converted to the equivalent form

$$\frac{1}{2}x^T(t_0)S(t_f,t_0;S_f)x(t_0) = \min_{v(\cdot)}\left\{\frac{1}{2}x^T(t_f)S_f x(t_f)\right. \tag{5.216}$$

$$\left. + \frac{1}{2}\int_{t_0}^{t_f}\left[x^T(t)\left(Q(t) - C^T(t)R^{-1}(t)C(t)\right)x(t) + v^T(t)R(t)v(t)\right]dt\right\},$$

where the cross term between $u(t)$ and $x(t)$ is eliminated. Since $R(t)$ is positive definite by Assumption 5.2.1 and by Assumption 5.6.1, S_f and $Q(t) - C^T(t)R^{-1}(t)C(t)$ are nonnegative definite, the cost must be nonnegative definite for all $x(t_0)$, i.e.,

$$x^T(t_0)S(t_f,t_0;S_f)x(t_0) \geq 0 \Rightarrow S(t_f,t_0;S_f) \geq 0, \tag{5.217}$$

regardless of t_0. Furthermore, since the original system is controllable with respect to u, then by Theorem 5.5.2, the new system

$$\dot{x}(t) = \left(A(t) - B(t)R^{-1}(t)C(t)\right)x(t) + B(t)v(t) \tag{5.218}$$

must be controllable with respect to $v(t)$. Therefore, the cost is bounded from above and below, even if t_0 goes to $-\infty$. ∎

The final two theorems deal with the monotonic and asymptotic behavior of the Riccati equation under Assumptions 5.2.1, 5.4.1, and 5.6.1. First, it is shown that when $S_f = 0$, $S(t_f,t;S_f)$ is a monotonically increasing function of t_f. Then, it is shown that $S^*(t_f, t; S_f)$ goes asymptotically to zero as a function of t as $t \to -\infty$ regardless of the boundary condition, S_f. This means that $S(t_f, t_0; S_f)$ and $\bar{S}(t_f, t_0)$ approach $S(t_f, t_0; 0)$ as the difference $t_f - t_0$ goes to infinity.

Theorem 5.6.3 *Given Assumptions* 5.2.1, 5.4.1, *and* 5.6.1,

$$S(t_f, t_0; 0) \geq S(t_1, t_0; 0) \quad \textit{for} \quad t_0 \leq t_1 \leq t_f. \tag{5.219}$$

Proof: The optimal cost criterion can be written as

$$\begin{aligned}
&\frac{1}{2} x^T(t_0) S(t_f, t_0; 0) x(t_0) \\
&\quad = \min_{u(\cdot)} \left\{ \frac{1}{2} \int_{t_0}^{t_f} \left[x^T(t) Q(t) x(t) + 2u^T(t) C(t) x(t) + u^T(t) R(t) u(t) \right] dt \right\} \\
&\quad = \min_{u(\cdot)} \left\{ \frac{1}{2} \int_{t_0}^{t_1} \left[x^T(t) Q(t) x(t) + 2u^T(t) C(t) x(t) \right. \right. \\
&\qquad\quad + \left. \left. u^T(t) R(t) u(t) \right] dt + \frac{1}{2} x^T(t_1) S(t_f, t_1; 0) x(t_1) \right\} \\
&\quad = \frac{1}{2} x^T(t_0) S(t_1, t_0; S(t_f, t_1; 0)) x(t_0).
\end{aligned} \tag{5.220}$$

From (5.210),

$$\begin{aligned}
&x^T(t_0) S(t_f, t_0; 0) x(t_0) \\
&\quad = x^T(t_0) S(t_1, t_0; S(t_f, t_1; 0)) x(t_0) \\
&\quad = x^T(t_0) \left[S(t_1, t_0; 0) + S^*\left(t_1, t_0; S(t_f, t_1; 0)\right) \right] x(t_0).
\end{aligned} \tag{5.221}$$

Since $S(t_f, t_1; 0) \geq 0$ by Theorem 5.6.2, then $S^*(t_1, t_0; S(t_f, t_1; 0)) \geq 0$. Therefore,

$$x_0^T S(t_f, t_0; 0) x_0 \geq x_0^T S(t_1, t_0; 0) x_0. \tag{5.222}$$

By Assumption 5.4.1, $x_0^T S(t_f, t_0; 0) x_0$ is bounded for all x_0 regardless of $t_0 \leq t_f$, implying

$$S(t_f, t_0; 0) \geq S(t_1, t_0; 0). \tag{5.223}$$

∎

In the second theorem we show that $S^*(t_f, t_0; S_f)$ goes to zero for all $S_f \geq 0$ as t_0 goes to $-\infty$ by requiring an observability assumption. If the cost with a terminal constraint is nonzero for all $x(t_0)$, $t_0 < t_f$, then $x^T(t_0)S(t_f, t_0; 0)x(t_0) > 0$ for all $x(t_0) \neq 0$. This condition will be shown to be guaranteed by ensuring that $y(t)$ given by

$$y(t) = N(t)x(t) \tag{5.224}$$

is observable where $N(t)$ is the square root of the matrix $\underline{Q} \triangleq Q(t) - C^T(t)R^{-1}(t)C(t)$, assumed to be nonnegative, and $y^T(t)y(t) = x^T(t)\underline{Q}(t)x(t)$.

Assumption 5.6.2 *The dynamic system*

$$\dot{x}(t) = A(t)x(t), \quad x(t_0) = x_0, \quad y(t) = N(t)x(t) \tag{5.225}$$

is completely observable on the interval $[t, t']$, *where* $t_0 < t < t' \leq t_f$.

If the system is completely observable, then the initial state can be determined as

$$x(t_0) = M^{-1}(t_0, t') \int_{t_0}^{t'} \Phi_A^T(t, t_0)N^T(t)y(t)dt, \tag{5.226}$$

where the observability Grammian matrix

$$M(t_0, t') = \int_{t_0}^{t'} \Phi_A^T(t, t_0)N^T(t)N(t)\Phi_A(t, t_0)dt \tag{5.227}$$

is invertible for all t' in the interval $t_0 < t' \leq t_f$. This means that

$$\int_{t_0}^{t'} x^T(t)\underline{Q}(t)x(t)dt = \int_{t_0}^{t'} y^T(t)y(t)dt = x_0^T M(t_0, t')x_0 > 0 \tag{5.228}$$

for all $x_0 \neq 0$, $t' \in (t_0, t_f]$ and $u(\cdot) = 0$. Therefore, $J(u(\cdot); x_0, t_0) > 0$.

Theorem 5.6.4 *Given Assumptions* 5.4.1, 5.6.1, *and* 5.6.2, *and* $\underline{A}(t)$ *defined in* (5.172) *with* $S_f = 0$, *the following hold:*

(a) $S(t_f, t_0; 0) > 0$ $\quad \forall\ t_0$ *in* $-\infty < t_0 < t_f$,
(b) $\dot{x}(t) = \underline{A}(t)x(t)$, $x(t_0) = x_0$ *is asymptotically stable*,
(c) $S^*(t_f, t_0; 0) \to 0$ *as* $t_0 \to -\infty\ \forall\ S_f \geq 0$.

Proof: From Theorem 5.6.2, $S(t_f, t_0; 0) \geq 0$ and bounded. By Assumption 5.6.2, using (5.216), for all $x_0 \neq 0$,

$$x_0^T S(t_f, t_0; 0) x_0 > 0 \tag{5.229}$$

for all t_0 in $-\infty < t_0 < t_f$. This results in $S(t_f, t_0; 0) > 0$. Let us now use the optimal cost function (5.229) as a Lyapunov function to determine if $x(t)$ in condition (b) is asymptotically stable. First, determine if the rate of change of the original cost function is negative definite. Therefore,

$$\begin{aligned} &\frac{d}{dt}\left[x^T(t) S(t_f, t; 0) x(t)\right] \\ &\quad = x^T(t)\left[\underline{A}^T(t) S(t_f, t; 0) + S(t_f, t; 0)\underline{A}(t) + \dot{S}(t_f, t; 0)\right] x(t). \end{aligned} \tag{5.230}$$

By using (5.78) with $S_f = 0$,

$$\begin{aligned} &\frac{d}{dt}\left[x^T(t) S(t_f, t; 0) x(t)\right] \\ &\quad = -x^T(t)\left[\underline{Q}(t) + S(t_f, t; 0) B(t) R(t)^{-1} B^T(t) S(t_f, t; 0)\right] x(t). \end{aligned} \tag{5.231}$$

Therefore, $\frac{d}{dt}\left[x(t)^T S(t_f, t; 0) x(t)\right] \leq 0$ and the optimal cost function (5.229) is a Lyapunov function for $t < t_f$. By integrating (5.231),

$$\begin{aligned} &x_0^T S(t_f, t_0; 0) x_0 - x^T(t_1) S(t_f, t_1; 0) x(t_1) \\ &\quad = \int_{t_0}^{t_1} x^T(t)\left[\underline{Q}(t) + S(t_f, t; 0) B(t) R^{-1}(t) B^T(t) S(t_f, t; 0)\right] x(t) dt, \end{aligned} \tag{5.232}$$

where $t_0 < t < t_f$ such that $S(t_f, t; 0) > 0$. If $\|x(t_1)\| \neq 0$, then $\|x(t)\| \neq 0$ for all $t \in [t_0, t_1]$, since $x(t)^T S(t_f, t; 0) x(t)$ is a Lyapunov function. Therefore, by Assumption 5.6.2, as $t_1 - t_0 \to \infty$ the right-hand side of (5.232) goes to ∞, implying that $x_0^T S(t_f, t_0; 0) x_0 \to \infty$. But this contradicts the fact that $S(t_f, t_0; 0) < \infty$

by Theorem 5.6.2 and x_0 is given and assumed finite. Therefore, $\|x(t_1)\| \to 0$ as $t_1 - t_0 \to \infty$ for any $S_f \geq 0$.

We now consider condition (c). In Remark 5.6.2 after Theorem 5.6.1, it is noted that $\bar{S}^*(t_f, t)$, defined in (5.213), satisfies the homogeneous Riccati equation (5.211). However, if the boundary conditions for $F(t_f, t_f)$ and $G(t_f, t_f)$ are chosen to be consistent with $S^*(t_f, t_f; S_f) = S_f$, then the solution and the behavior of $S(t_f, t; S_f)$ can be determined from $F(t_f, t)$ and $G(t_f, t)$. By writing

$$S_f = KK^T, \tag{5.233}$$

then

$$F(t_f, t_f) = K, \quad G(t_f, t_f) = -I. \tag{5.234}$$

Note that $F(t_f, t)$ satisfies, from (5.161), the differential equation

$$\dot{F}(t_f, t) = -F(t_f, t)\underline{A}(t)^T, \quad F(t_f, t_f) = K. \tag{5.235}$$

From the conditions (b) of the theorem, $F(t_f, t)$ is stable and approaches zero as t goes to $-\infty$. From (5.164) and (5.234), the evaluation of $G(t_f, t)$ must always be negative definite for $t \leq t_f$, implying that $G(t_f, t)^{-1}$ exist for all $t \leq t_f$. Since $F(t_f, t) \to 0$ as $t \to -\infty$, $S^*(t_f, t; S_f) \to 0$ as $t \to -\infty$. ∎

Remark 5.6.3 *Note that* $S(t_f, t; S_f) \to S(t_f, t; 0)$ *as* $t \to -\infty$ *for all* $S_f \geq 0$.

Remark 5.6.4 *The assumptions of controllability and observability are stronger conditions than are usually needed. For example, if certain states are not controllable but naturally decay, then stabilizability of* $x(t)$ *is sufficient for condition* (b) *of Theorem* 5.6.4 *to still hold.*

5.7 LQ Regulator Problem

The LQ control problem is restricted in this section to a constant coefficient dynamic system and cost criterion. Furthermore, the time interval over which the cost criterion is to be minimized is assumed to be infinite. As might be suspected by the previous results, a linear constant gain controller results from this restricted formulation. This specialized problem is sometimes referred to as the linear quadratic regulator (LQR) problem. linear quadratic regulator problem

The optimal control problem of Section 5.2 is specialized to the linear regulator problem by requiring that t_f be infinite and A, B, Q, C, and R are all constant matrices.

Theorem 5.7.1 *For the LQR problem, given Assumptions* 5.4.1, 5.6.1, *and* 5.6.2, *there is a unique, symmetric, positive-definite solution, S, to the algebraic Riccati equation (ARE)*

$$\begin{aligned}&(A - BR^{-1}C)S + S(A - BR^{-1}C)^T + (Q - C^T R^{-1}C)\\&\quad - SBR^{-1}B^T S = 0\end{aligned} \tag{5.236}$$

such that $(A - BR^{-1}C - BR^{-1}B^T S)$ has only eigenvalues with negative real parts.

Proof: From Theorem 5.6.3, $S(t_f, t_0; 0)$ is monotonic in t_f. Since the parameters are not time dependent, then $S(t_f, t_0; 0)$ depends only upon $t_f - t_0$ and is monotonically increasing with respect to $t_f - t_0$. Since from Theorem 5.6.2, $S(t_f, t_0; 0)$ is bounded for all $t_f - t_0$, then as $t_f - t_0 \to \infty$, $S(t_f, t_0; 0)$ reaches an upper limit of S. As $S(t_f, t_0; 0)$ approaches S, $\dot{S}(t_f, t; 0)$ approaches zero, implying that S must satisfy the ARE (5.236). That is, for some $\Delta > 0$,

$$\begin{aligned}S(t_f, t_0 - \Delta; 0) &= S(t_f - \Delta, t_0 - \Delta; S(t_f, t_f - \Delta; 0))\\&= S(t_f, t_0; S(t_0, t_0 - \Delta; 0)),\end{aligned} \tag{5.237}$$

where the time invariance of the system is used to shift the time, i.e., $S(t_f, t_f - \Delta; 0) = S(t_0, t_0 - \Delta; 0)$ and $t_f - \Delta$, $t_0 - \Delta$ become t_f, t_0. Continuity of the solution with respect to the initial conditions implies that as $\Delta \to \infty$, $S(t_0, t_0 - \Delta; 0)$ and $S(t_f, t_0 - \Delta; 0)$ go to S such that (5.237) becomes

$$S = S(t_f, t_0; S), \tag{5.238}$$

and S is a fixed-point solution to the autonomous (time-invariant) Riccati equation. Furthermore, by condition (c) of Theorem 5.6.4, $S(t_f, t_0; S_f)$ approaches the same limit regardless of $S_f \geq 0$ and, therefore, S is unique. By conditions (a) and (b) of Theorem 5.6.4, S is positive-definite and x is asymptotically stable. Since S is a constant, this implies that the eigenvalues of the constant matrix

$$\underline{A} \triangleq A - BR^{-1}C - BR^{-1}B^T S \tag{5.239}$$

have only negative real parts. ▮

The relationship between the Hamiltonian matrix, H, of (5.48) and the feedback dynamic matrix $\underline{A}$ of (5.239) is vividly obtained by using the canonical transformation introduced in Section 5.4.3. First, additional properties of Hamiltonian systems are obtained for the constant Hamiltonian matrix by rewriting (5.63) as

$$H = -\underline{J}^T H^T \underline{J}. \tag{5.240}$$

The characteristic equations for H and $-H$ can be obtained by subtracting λI from both sides of (5.240) and taking the determinant as

$$\begin{aligned} \det(H - \lambda I) &= \det(-J^T H^T J - \lambda I) \\ &= \det(-J^T H^T J - \lambda J^T J) \\ &= \det J^T \det(-H - \lambda I) \det J \\ &= \det(-H - \lambda I). \end{aligned} \tag{5.241}$$

Since the characteristic equations for H and $-H$ are equal, the eigenvalues of the $2n \times 2n$ matrix H are not only symmetric about the real axis but also about the imaginary axis. If n eigenvalues are λ_i, $i = 1, \ldots, n$, then the remaining n eigenvalues are

$$\lambda_{i+n} = -\lambda_i, \quad i = 1, \ldots, n. \tag{5.242}$$

From (5.242), it is seen that there are just as many stable eigenvalues as unstable ones. Furthermore, there is a question as to how many eigenvalues lie on the imaginary axis.

To better understand the spectral content of H, the canonical transformation of (5.84) is used with the steady state S, which is the solution to the ARE (5.236). By using this canonical transformation, similar to (5.84), the transformed Hamiltonian matrix H is

$$LHL^{-1} \triangleq \bar{H} = \begin{bmatrix} \underline{A} & -BR^{-1}B^T \\ 0 & -\underline{A}^T \end{bmatrix}. \tag{5.243}$$

This form is particularly interesting because $\underline{A}$ and $-\underline{A}^T$ contain all the spectral information of both $\bar{H}$ and H since L is a similarity transformation. Note that from Theorem 5.7.1, the real parts of the eigenvalues of $\underline{A}$ are negative. Therefore, the feedback dynamic matrix $\underline{A}$ with $S > 0$ contains all the left-half plane poles in H. The numerous solutions to the ARE, in which it will be shown that $S > 0$ is only one of many solutions, will give various groupings of the eigenvalues of the Hamiltonian matrix.

If the Hamiltonian matrix has no eigenvalues on the imaginary axis, then there is at most one solution to the matrix Riccati equation which will decompose the matrix such that the eigenvalues of $\underline{A}$ have only negative real parts. This will be the case even if we are not restricted to Assumptions 5.4.1, 5.6.1, and 5.6.2. The following theorem from Brockett [8] demonstrates directly that if the real parts of the eigenvalues of $\underline{A}$ are negative, then S is unique.

Theorem 5.7.2 *For the LQR problem, there is at most one symmetric solution of $SA + A^T S - SBR^{-1}B^T S + Q = 0$ having the property that the eigenvalues of $A - BR^{-1}B^T S$ have only negative real parts.*

Proof: Assume, to the contrary, that there are two symmetric solutions, S_1 and S_2, such that both the eigenvalues of $\underline{A}_1 = A - BR^{-1}B^T S_1$ and $\underline{A}_2 = A - BR^{-1}B^T S_2$ have only negative real parts. By proceeding as in (5.94) to (5.96), the cost can be written as

$$\begin{aligned} &J\left(u(\cdot); x(t_0), t_0\right) \qquad (5.244)\\ &= \frac{1}{2}\left[x^T(t_0)S_1 x(t_0) - x^T(t_f)S_1 x(t_f)\right]\\ &+ \int_{t_0}^{t_f} \frac{1}{2}\left[u(t) + R^{-1}\left(C + B^T S_1\right)x(t)\right]^T R\left[u(t) + R^{-1}\left(C + B^T S_1\right)x(t)\right] dt\\ &= \frac{1}{2}\left[x^T(t_0)S_2 x(t_0) - x^T(t_f)S_2 x(t_f)\right]\\ &+ \int_{t_0}^{t_f} \frac{1}{2}\left[u(t) + R^{-1}\left(C + B^T S_2\right)x(t)\right]^T R\left[u(t) + R^{-1}\left(C + B^T S_2\right)x(t)\right] dt. \end{aligned}$$

Since S_1 and S_2 are symmetric and $S_1 \neq S_2$, then there exists an x_0 such that $x_0^T S_1 x_0 \neq x_0^T S_2 x_0$. Suppose that $x_0^T S_1 x_0 \geq x_0^T S_2 x_0$ and let $u(t) = -R^{-1}(C + B^T S_2)x(t)$. By taking the limits as t_f goes to infinity means $x(t_f)$ goes to zero by the assumption that $\underline{A}_1$ and $\underline{A}_2$ are stable, and the cost can be written as

$$\begin{aligned} J\left(u(\cdot); x(t_0), t_0\right) &= \frac{1}{2}x^T(t_0)S_2 x(t_0) = \frac{1}{2}x^T(t_0)S_1 x(t_0) \qquad (5.245)\\ &+ \int_{t_0}^{\infty} \frac{1}{2}\left[B^T\left(S_1 - S_2\right)x(t)\right]^T R^{-1}\left[B^T\left(S_1 - S_2\right)x(t)\right] dt, \end{aligned}$$

which contradicts the hypothesis that $x_0^T S_1 x_0 \geq x_0^T S_2 x_0$ and, therefore, the hypothesis that there can be two distinct solutions to the ARE, which both produce stable dynamic matrices $\underline{A}_1$ and $\underline{A}_2$. ▮

Remark 5.7.1 *Since Assumptions* 5.4.1, 5.6.1, *and* 5.6.2 *are not required, then the solution to the ARE is not necessarily nonnegative definite. Furthermore, requiring Assumptions* 5.4.1, 5.6.1, *and* 5.6.2 *implies that as* $t \to -\infty$, $\bar{S}(t_f, t) \to S(t_f, t; 0)$. *This is not generally the case. From Theorem* 5.5.3, $\bar{S}(t_f, t) \geq S(t_f, t; 0)$. *It can well occur that* $S(t_f, t; 0)$ *can have a finite escape time, whereas* $\bar{S}(t_f, t; 0)$ *remains bounded even as* $t \to -\infty$. *For the regulator problem, the unconstrained terminal control problem may not have a finite cost. However, by requiring that* $\lim_{t_f \to \infty} x(t_f) \to 0$, *the terminally constrained problem may have a finite cost where* $\lim_{t \to -\infty} \bar{S}(t_f, t) \to S$.

Remark 5.7.2 *It is clear from the canonical transformation of the Hamiltonian matrix* H *into* $\bar{H}$ *of* (5.243) *that a necessary condition for all the eigenvalues of* $\underline{A}$ *to have negative real parts is for* H *to have no eigenvalues which have zero real parts.*

5.8 Necessary and Sufficient Conditions for Free Terminal Time

In this section necessary and sufficient conditions for optimality of the second variation to be positive and strongly positive for the terminally constrained, free terminal time problem are determined. We return to Section 5.1, where in (5.10) the second variational augmented cost criterion $\delta^2 \hat{J}$, with augmented variation in the terminal constraints and free terminal time, is given. In (5.10) the linear dynamics in z are adjoined by the Lagrange multiplier $\tilde{\lambda}$ in the integrand and the terminal constraints are adjoined by $\tilde{\nu}$ in the quadratic terminal function. For consistency of notation, the variables given in Section 5.2 and using Remark 5.2.1, where the augmented cost (5.10) is considered rather than (5.15) and for this free terminal time problem, the additional notation $\tilde{D}^T \triangleq \frac{d\psi}{dt}$, $E \triangleq \Omega_x$, and $\tilde{E} \triangleq \frac{d\Omega}{dt}$ is used. In particular, (5.24) is extended for free terminal time, having integrated $\lambda^T(t)x(t)$ by parts and using the

Hamiltonian given in (5.23), as

$$\begin{aligned}
&\hat{J}\left(u(\cdot),\Delta;x_0,\lambda(\cdot),\nu\right) \\
&= \int_{t_0}^{t_f^o}\left[H(x(t),\lambda(t)u(t),t)+\dot{\lambda}^T(t)x(t)\right]dt+\lambda^T(t_0)x_0-\lambda^T(t_f^o)x(t_f^o) \\
&\quad +\frac{1}{2}\left[\begin{array}{ccc} x(t_f^o)^T & \nu^T & \Delta \end{array}\right]\left[\begin{array}{ccc} S_f & D^T & E^T \\ D & 0 & \tilde{D}^T \\ E & \tilde{D} & \tilde{E} \end{array}\right]\left[\begin{array}{c} x(t_f^o) \\ \nu \\ \Delta \end{array}\right],
\end{aligned} \tag{5.246}$$

where $\tilde{\lambda}$ and $\tilde{\nu}$ have been replaced by λ and ν, respectively. The notation is now consistent with that given for the LQ problem except for the inclusion of Δ, the variation in the terminal time, which is a control parameter entering only in the terminal function.

To determine the first-order necessary conditions for variations in $\hat{J}$ to be non-negative, variations in the control are made as $\delta u(t) = u(t) - u^o(t)$ and produces a variation in $x(t)$ as $\delta x(t) = x(t) - x^o(t)$. Furthermore, a variation in Δ must also be considered as $\delta\Delta = \Delta - \Delta^o$. The change in $\hat{J}$ is made using these variations and thereby the expansion of (5.28) is extended for free terminal time as

$$\begin{aligned}
\Delta\hat{J} &\stackrel{\triangle}{=} \Delta\hat{J}(u(\cdot),u^o(\cdot),\Delta,\Delta^o;x_0,\lambda(\cdot),\nu) \\
&= \hat{J}\left(u(\cdot),\Delta;x_0,\lambda(\cdot),\nu\right)-\hat{J}\left(u^o(\cdot),\Delta^o;x_0,\lambda(\cdot),\nu\right) \\
&= \int_{t_0}^{t_f^o}\Big[x^{o^T}(t)Q(t)\delta x(t)+u^{o^T}(t)C(t)\delta x(t)+x^{o^T}(t)C^T(t)\delta u(t) \\
&\qquad +\frac{1}{2}\delta u^T(t)R(t)\delta u(t)+u^{o^T}(t)R(t)\delta u(t)+\dot{\lambda}^T(t)\delta x(t) \\
&\qquad +\lambda^T(t)A(t)\delta x(t)+\lambda^T(t)B(t)\delta u(t)\Big]dt \\
&\quad +(-\lambda^T(t_f^o)+x^{o^T}(t_f^o)S_f+\nu^T D+\Delta^o E)\delta x(t_f) \\
&\quad +(x^{o^T}(t_f^o)E^T+\nu^T\tilde{D}^T+\Delta^o\tilde{E})\delta\Delta+\int_{t_0}^{t_f^o}\mathcal{O}(t;\varepsilon)dt+\mathcal{O}(\varepsilon),
\end{aligned} \tag{5.247}$$

where variations of $\lambda(t)$ and ν multiply the dynamic constraints and the terminal constraint, which we assume are satisfied. Following a procedure similar to that which produced Theorem 5.3.1, we summarize these results for the free terminal time problem as Theorem 5.8.1.

Theorem 5.8.1 *Suppose that Assumptions* 5.2.1, 5.2.2, *and* 5.3.1 *are satisfied. Then the necessary conditions for* ΔJ *to be nonnegative to first order for strong perturbations in the control* (5.26) *are*

$$\begin{aligned}
\dot{x}^o(t) &= A(t)x^o(t) + B(t)u^o(t), \quad x^o(t_0) = x_0, && (5.248)\\
\dot{\lambda}(t) &= -A^T(t)\lambda(t) - Q(t)x^o(t) - C^T(t)u^o(t), && (5.249)\\
\lambda(t_f^o) &= S_f x^o(t_f^o) + D^T\nu + E^T\Delta^o, && (5.250)\\
0 &= Dx^o(t_f^o) + \tilde{D}^T\Delta^o, && (5.251)\\
0 &= Ex^o(t_f^o) + \tilde{D}\nu + \tilde{E}\Delta^o, && (5.252)\\
0 &= R(t)u^o(t) + C(t)x^o(t) + B^T(t)\lambda(t). && (5.253)
\end{aligned}$$

For free terminal time, the boundary condition for the Lagrange multiplier is (5.250), the terminal constraint is (5.251), and the transversality condition is (5.252).

Using the transition matrix of (5.50) to relate the initial and terminal values of (x, λ) in the above equations (5.250, 5.251, 5.252) and simplifying the notation by using (5.72), we obtain a new expression similar to (5.154) except that the variation in terminal time is included. The generalization of (5.154) for the free terminal time problem has the form

$$\begin{bmatrix} \lambda(t_0) \\ 0 \\ 0 \end{bmatrix} = \begin{bmatrix} -\bar{\Phi}_{22}^{-1}(t_f^o, t_0)\bar{\Phi}_{21}(t_f^o, t_0) & \bar{\Phi}_{22}^{-1}(t_f^o, t_0)D^T & \bar{\Phi}_{22}^{-1}(t_f^o, t_0)E^T \\ D\bar{\Phi}_{22}^{-T}(t_f^o, t_0) & D\bar{\Phi}_{12}(t_f^o, t_0)\bar{\Phi}_{22}^{-1}(t_f^o, t_0)D^T & D\bar{\Phi}_{12}(t_f^o, t_0)\bar{\Phi}_{22}^{-1}(t_f^o, t_0)E^T + \tilde{D}^T \\ E\bar{\Phi}_{22}^{-T}(t_f^o, t_0) & E\bar{\Phi}_{12}(t_f^o, t_0)\bar{\Phi}_{22}^{-1}(t_f^o, t_0)D^T + \tilde{D} & E\bar{\Phi}_{12}(t_f^o, t_0)\bar{\Phi}_{22}^{-1}(t_f^o, t_0)E^T + \tilde{E} \end{bmatrix} \begin{bmatrix} x_0 \\ \nu \\ \Delta^o \end{bmatrix}. \quad (5.254)$$

We now make the following identifications with $t = t_0$ as done in (5.155), (5.156), and (5.157) with the elements in (5.254), yielding the symmetric form

$$\begin{bmatrix} \lambda(t) \\ 0 \\ 0 \end{bmatrix} = \begin{bmatrix} S(t_f^o, t; S_f) & F^T(t_f^o, t) & m^T(t_f^o, t) \\ F(t_f^o, t) & G(t_f^o, t) & n^T(t_f^o, t) \\ m(t_f^o, t) & n(t_f^o, t) & s(t_f^o, t) \end{bmatrix} \begin{bmatrix} x(t) \\ \nu \\ \Delta^o \end{bmatrix}. \tag{5.255}$$

The differential properties of $-\bar{\Phi}_{22}^{-1}(t_f^o, t)\bar{\Phi}_{21}(t_f^o, t)$, $\bar{\Phi}_{22}^{-1}(t_f^o, t)$, and $\bar{\Phi}_{12}(t_f^o, t)$ $\bar{\Phi}_{22}^{-1}(t_f^o, t)$ are given in Theorem 5.4.1, (5.160), and (5.163), and therefore the differential equations for $S(t_f^o, t, S_f)$, $F(t_f^o, t)$, and $G(t_f^o, t)$ are given in (5.78), (5.161), and (5.164). The dynamics of $m(t_f^o, t)$, $n(t_f^o, t)$, and $s(t_f^o, t)$ are determined from (5.160) and (5.163) as

$$\begin{aligned} \frac{d}{dt}\left[m^T(t_f^o, t)\right] &= -\Big[\big(A(t) - B(t)R(t)^{-1}C(t)\big)^T - S(t_f^o, t; S_f)B(t)R(t)^{-1}B^T(t)\Big] \\ &\qquad \times m^T(t_f^o, t), \quad m(t_f^o, t_f^o) = E, &(5.256) \\ \frac{d}{dt}\left[n^T(t_f^o, t)\right] &= F(t_f^o, t)B(t)R(t)^{-1}B^T(t)m^T(t_f^o, t), \quad n(t_f^o, t_f^o) = \tilde{D}, &(5.257) \\ \frac{d}{dt}\left[s(t_f^o, t)\right] &= m(t_f^o, t)B(t)R(t)^{-1}B^T(t)m^T(t_f^o, t), \quad s(t_f^o, t_f^o) = \tilde{E}. &(5.258) \end{aligned}$$

This approach is sometimes called the sweep method because the boundary conditions at $t = t_f^o$ are swept backward to the initial time. For free terminal time the quadratic cost can be shown to reduce to a form

$$J\left(u^o(\cdot); x(t'), t'\right) = \frac{1}{2}x^T(t')\tilde{S}(t_f^o, t')x(t'), \tag{5.259}$$

where t' is close enough to t_f^o such that $\tilde{S}(t_f^o, t')$ exists. $\tilde{S}(t_f^o, t')$ is determined by relating $\lambda(t')$ to $x(t')$ as

$$\lambda(t') = \tilde{S}(t_f^o, t')x(t'), \tag{5.260}$$

where ν and Δ^o are eliminated in (5.255) in terms of $x(t)$ as

$$-\begin{bmatrix} G(t_f^o,t) & n^T(t_f^o,t) \\ n(t_f^o,t) & s(t_f^o,t) \end{bmatrix} \begin{bmatrix} \nu \\ \Delta^o \end{bmatrix} = \begin{bmatrix} F(t_f^o,t) \\ m(t_f^o,t) \end{bmatrix} x(t)$$

$$\Rightarrow \begin{bmatrix} \nu \\ \Delta^o \end{bmatrix} = -\begin{bmatrix} G(t_f^o,t) & n^T(t_f^o,t) \\ n(t_f^o,t) & s(t_f^o,t) \end{bmatrix}^{-1} \begin{bmatrix} F(t_f^o,t) \\ m(t_f^o,t) \end{bmatrix} x(t). \tag{5.261}$$

Substituting back into (5.255) results in (5.260), where

$$\tilde{S}(t_f^o,t) = \Bigg[S(t_f^o,t;S_f) - \begin{bmatrix} F(t_f^o,t)^T & m^T(t_f^o,t) \end{bmatrix} \begin{bmatrix} G(t_f^o,t) & n^T(t_f^o,t) \\ n(t_f^o,t) & s(t_f^o,t) \end{bmatrix}^{-1} \times \begin{bmatrix} F(t_f^o,t) \\ m(t_f^o,t) \end{bmatrix} \Bigg]. \tag{5.262}$$

Remark 5.8.1 *The proof for positivity of the second variational cost remains the same as given in Theorem* 5.5.3. *Furthermore,* $\tilde{S}(t_f^o,t)$ *satisfies the same Riccati equation as* $S(t_f^o,t;S_f)$*, given in Theorem* 5.4.1. *Similarly, the feedback control with free terminal time and terminal constraints is given by* (5.168)*, where* $\bar{S}(t_f,t)$ *is replaced by* $\tilde{S}(t_f^o,t)$*. For the problem to be normal the inverse in* (5.262) *must exist.*

Remark 5.8.2 *Construction of* $\tilde{S}(t_f^o,t)$ *by* (5.262) *using its component parts may not be possible since* $S(t_f,t;S_f)$ *may not exist, whereas* $\tilde{S}(t_f^o,t)$ *can exist.*

Remark 5.8.3 *Theorem* 5.5.3 *can be extended to show that* $J\left(u(\cdot),t_f^o;0,t_0\right)$ *is strongly positive definite by a proof similar to that used in Theorem* 5.4.3.

Remark 5.8.4 *Note that the optimal value function for the LQ problem with terminal constraints is given as*

$$V(x(t),t) = \frac{1}{2}x^T(t)\tilde{S}(t_f^o,t)x(t). \tag{5.263}$$

This satisfies the Hamilton–Jacobi–Bellman equation given in Theorem 4.7.1.

Example 5.8.1 (Second variation with variable terminal time) *The objective of this example is to determine the first- and second-order necessary conditions for a free terminal time optimal control problem. The problem statement is as follows: find the control scheme which minimizes*

$$J = t_f \triangleq \phi(x_f, t_f) \tag{5.264}$$

with the dynamic system equations

$$\dot{x} = \begin{bmatrix} \dot{x}_1 \\ \dot{x}_2 \end{bmatrix} = \begin{bmatrix} v\cos\beta \\ v\sin\beta \end{bmatrix} \triangleq f, \quad x(t_0) = \begin{bmatrix} x_{10} \\ x_{20} \end{bmatrix} \triangleq x_0 \tag{5.265}$$

and the terminal boundary condition

$$\psi(x_f, t_f) = \begin{bmatrix} x_{1f} \\ x_{2f} \end{bmatrix} \triangleq x_f = 0, \tag{5.266}$$

where v is constant and β is the control variable.

The augmented performance index is

$$J = t_f + \nu^T\psi + \int_{t_0}^{t_f} \left[H - \lambda^T\dot{x}\right] dt, \tag{5.267}$$

where

$$H \triangleq \lambda^T f = \lambda_1 v\cos\beta + \lambda_2 v\sin\beta, \tag{5.268}$$

$$\tilde{\phi}(t_f, x_f, \nu) = t_f + \nu^T x_f. \tag{5.269}$$

First-order necessary conditions from Chapter 4 are

$$\dot{x} = H_\lambda^T = f, x(t_0) = x_0, \tag{5.270}$$

$$\dot{\lambda} = -H_x^T = 0,\ \lambda(t_f) = \tilde{\phi}_{x_f}^T = \nu \Rightarrow \lambda(t) = \nu, \tag{5.271}$$

$$0 = H_u = H_\beta = \lambda_2 v\cos\beta - \lambda_1 v\sin\beta, \tag{5.272}$$

and the transversality condition is

$$0 = H(t_f) + \tilde{\phi}_{t_f} = \lambda_1 v\cos\beta_f + \lambda_2 v\sin\beta_f + 1. \tag{5.273}$$

From the optimality condition (5.272)

$$\lambda_2 \cos\beta(t) = \lambda_1 \sin\beta(t) \;\Rightarrow\; \tan\beta(t) = \frac{\lambda_2}{\lambda_1} = \text{constant}$$
$$\Rightarrow\; \beta^o(t) = \text{constant}. \tag{5.274}$$

Integrating the dynamics (5.265) *while keeping* $\beta^o(t) = \beta^o = \text{constant}$, *we obtain*

$$\begin{aligned} x_1(t) &= x_{10} + (t - t_0)v\cos\beta^o, \\ x_2(t) &= x_{20} + (t - t_0)v\sin\beta^o. \end{aligned} \tag{5.275}$$

Substituting (5.275) *into* (5.266), *we obtain*

$$\begin{aligned} x_{10} + (t_f^o - t_0)v\cos\beta^o &= 0, \\ x_{20} + (t_f^o - t_0)v\sin\beta^o &= 0. \end{aligned} \tag{5.276}$$

Solving (5.276) *for* β^o, *we obtain*

$$\tan\beta^o = \frac{x_{20}}{x_{10}} \Rightarrow \beta^o = \tan^{-1}\left(\frac{x_{20}}{x_{10}}\right). \tag{5.277}$$

Next we determine t_f^o. *If we restrict* $x_{10} > 0$ *and* $x_{20} > 0$, *then* β *satisfies* $\pi < \beta^o < \frac{3\pi}{2}$ *and* (5.277) *has only one solution. Thus*

$$\cos\beta^o = \frac{-x_{10}}{\sqrt{x_{10}^2 + x_{20}^2}}, \qquad \sin\beta^o = \frac{-x_{20}}{\sqrt{x_{10}^2 + x_{20}^2}}. \tag{5.278}$$

Therefore, from (5.276)

$$t_f^o = t_0 + \frac{\sqrt{x_{10}^2 + x_{20}^2}}{v}. \tag{5.279}$$

Substituting (5.278) *into* (5.274) *and the transversality condition* (5.273), *we obtain*

$$\begin{aligned} \lambda_2(-x_{10}) &= \lambda_1(-x_{20}), \\ 0 &= \frac{-\lambda_1 x_{10}}{\sqrt{x_{10}^2 + x_{20}^2}} + \frac{-\lambda_2 x_{20}}{\sqrt{x_{10}^2 + x_{20}^2}} + \frac{1}{v}. \end{aligned} \tag{5.280}$$

Solving these equations for λ_1 *and* λ_2*, we obtain*

$$\lambda_1 = \nu_1 = \frac{x_{10}}{v\sqrt{x_{10}^2 + x_{20}^2}}, \qquad \lambda_2 = \nu_2 = \frac{x_{20}}{v\sqrt{x_{10}^2 + x_{20}^2}}. \tag{5.281}$$

Next, the second variation necessary and sufficient conditions are to be developed. The necessary conditions for the variation in the cost criterion to be nonnegative are given in Theorem 5.8.1*. For the example they are*

$$\begin{aligned}
\dot{x} &= Bu, \quad \text{where } B = f_\beta = \begin{bmatrix} f_{1\beta} \\ f_{2\beta} \end{bmatrix} = \begin{bmatrix} -v\sin\beta^o \\ v\cos\beta^o \end{bmatrix}, && (5.282)\\
\dot{\lambda} &= 0,\ \lambda(t_f^o) = D\nu = I\nu \Rightarrow \lambda(t) = \nu, && (5.283)\\
0 &= Dx(t_f^o) + \tilde{D}^T\Delta^o = x(t_f^o) + f\Delta^o, && (5.284)\\
0 &= \tilde{D}\nu = f^T\nu, \quad \text{where } E = 0 \text{ and } \tilde{E} = 0, && (5.285)\\
0 &= v(\lambda_2\cos\beta^o - \lambda_1\sin\beta^o) + u, && (5.286)\\
&\therefore\ u = -v(\lambda_2\cos\beta^o - \lambda_1\sin\beta^o). && (5.287)\\
H_{\beta\beta} &= R = 1. && (5.288)
\end{aligned}$$

Therefore, the terminal boundary conditions for (5.255) *are given by* (5.283)*,* (5.284)*, and* (5.285) *as*

$$\begin{bmatrix} \lambda(t_f^o) \\ 0 \\ 0 \end{bmatrix} = \begin{bmatrix} 0 & I & 0 \\ I & 0 & f \\ 0 & f^T & 0 \end{bmatrix} \begin{bmatrix} x(t_f^o) \\ \nu \\ \Delta^o \end{bmatrix}. \tag{5.289}$$

From (5.255) *the differential equations* (5.78), (5.161), (5.164), (5.256), (5.257)*, and* (5.258) *are solved, where* $A = 0$*,* $Q = 0$*,* $C = 0$*, and* B *and* R *are given above. Therefore,* $S(t_f^o, t; S_f) = 0$*,* $F(t_f^o, t) = I$*,* $m(t_f^o, t) = 0$*,* $G(t_f^o, t) = f_\beta f_\beta^T \Delta t$*,* $n(t_f^o, t) = f^T$*, and* $s(t_f^o, t) = 0$*, where* $\Delta t = t_f - t$*. Therefore,* (5.255) *reduces to*

$$\begin{bmatrix} \lambda(t) \\ 0 \\ 0 \end{bmatrix} = \begin{bmatrix} 0 & I & 0 \\ I & f_\beta f_\beta^T \Delta t & f \\ 0 & f^T & 0 \end{bmatrix} \begin{bmatrix} x(t) \\ \nu \\ \Delta^o \end{bmatrix}. \tag{5.290}$$

From (5.290)

$$\begin{bmatrix} f_\beta f_\beta^T \Delta t & f \\ f^T & 0 \end{bmatrix} \begin{bmatrix} \nu \\ \Delta^o \end{bmatrix} = \begin{bmatrix} -I \\ 0 \end{bmatrix} x(t). \tag{5.291}$$

The coefficient matrix of (5.291) *is invertible for* $\Delta t > 0$, *i.e., the determinant is* $-v^4 \Delta t$, *so that*

$$\begin{aligned} \begin{bmatrix} \nu \\ \Delta^o \end{bmatrix} &= -\begin{bmatrix} f_\beta f_\beta^T \Delta t & f \\ f^T & 0 \end{bmatrix}^{-1} \begin{bmatrix} I \\ 0 \end{bmatrix} x(t) \\ &= -\begin{bmatrix} \dfrac{\sin^2 \beta^o}{v^2 \Delta t} & \dfrac{-\sin \beta^o \cos \beta^o}{v^2 \Delta t} \\ \dfrac{-\cos \beta^o \sin \beta^o}{v^2 \Delta t} & \dfrac{\cos^2 \beta^o}{v^2 \Delta t} \\ \dfrac{\cos \beta^o}{v} & \dfrac{\sin \beta^o}{v} \end{bmatrix} x(t). \end{aligned} \tag{5.292}$$

Substitution of (5.292) *into* (5.290) *gives* $\lambda(t) = \tilde{S}(t_f, t) x(t)$, *where*

$$\tilde{S}(t_f^o, t) = -\frac{1}{v^2 \Delta t} \begin{bmatrix} \sin \beta^o \\ -\cos \beta^o \end{bmatrix} \begin{bmatrix} \sin \beta^o & -\cos \beta^o \end{bmatrix}. \tag{5.293}$$

$\tilde{S}(t_f^o, t)$ *is well behaved for all* $t < t_f^o$, *where* $\Delta t > 0$. *Therefore, the extremal path satisfies the condition for positivity and is a local minimum. Note that* $\tilde{S}(t_f^o, t) \geq 0$ *is only positive semidefinite. This is because perturbation orthogonal to the extremal path does not affect the cost.*

5.9 Summary

A consistent theory is given for a rather general formulation of the LQ problem. The theory has emphasized the time-varying formulation of the LQ problem with linear terminal constraints. The relationship between the transition matrix of the Hamiltonian system and the solution to the matrix Riccati differential equation is vividly shown using the symplectic property of Hamiltonian systems. Initially, no requirement was placed on the state weighting in the cost function, although the control weighting was assumed to be positive definite. By using this cost criterion, the existence of the solution to the Riccati differential equation is required for the cost criterion to be positive definite and strongly positive definite. If this problem were interpreted as the accessory problem in the calculus of variations [6], then the requirement that the cost criterion be positive definite is not enough. As shown in Section 5.4.5 and [22, 4] the cost criterion is required to be strongly positive to ensure that the second variation dominates over higher-order terms.

The second-order necessary and sufficiency conditions are for extremal paths which satisfy weak or strong first-order conditions. However, it is assumed that there are no discontinuous changes in the optimal control history. If there are discontinuities, then the variation of the control about the discontinuity must be a strong variation. For example, see the Bushaw problem in Sections 4.6.1 and 4.7.1, where the control is bang-bang. This is not included in the current theory. In fact, the development of second-order conditions is done by converting the problem from explicitly considering strong variations to assuming that the times that the control switches are control parameters. With respect to these control parameters, the local optimality of the path is determined. First-order optimality of the cost criterion with respect to the switch time is zero and corresponds to the switch condition for the control with respect to the Hamiltonian. For example, in the Bushaw problem, the switch occurs when H_u goes through zero. Second-order optimality of the cost criterion with respect to the switch time can be found in [32] and [20]. Additional extensions of the LQ theory to several classes of nonlinear systems and other advanced topics are given in [29] and [31].

Problems

1. If two square matricies are symplectic, show that their product is symplectic.

2. A matrix H is "Hamiltonian" if it satisfies

$$J^{-1}H^T J = -H, \tag{5.294}$$

where

$$J \triangleq \begin{bmatrix} 0 & I \\ -I & 0 \end{bmatrix} \tag{5.295}$$

is the fundamental symplectic matrix. Show that the matrix $U^{-1}HU$ is also Hamiltonian, where U is a symplectic matrix.

3. Minimize the performance index

$$J = \int_0^{t_f} \left[cxu + \frac{1}{2}u^2 \right] dt$$

with respect to $u(\cdot)$ and subject to

$$\dot{x} = u, \qquad x(0) = 1.$$

(a) What are the first- and second-order necessary conditions for optimality? Don't bother to solve them.

(b) For $c = 1$ what are the values of t_f for which the second variation is nonnegative?

(c) For $c = -1$ what are the values of t_f for which the second variation is nonnegative?

4. Consider the problem of minimizing with respect to the control $u(\cdot) \in \mathcal{U}_{TB}$ the cost criterion

$$J = \int_0^{t_f} (a^T x + u^T R u) dt, \quad R > 0,$$

subject to

$$\dot{x} = Ax + Bu; \quad x(0) = x_0 \text{ (given)}, \quad x(t_f) = 0.$$

(a) By using the results of the accessory minimum problem, show that the extremal is locally minimizing.

(b) Relate the results of the second variation to controllability.

(c) Show that the second variation is strongly positive.

(d) Show that by extremizing the cost with respect to the Lagrange multiplier associated with the terminal constraints, the multipliers maximize the performance index.

5. Show that $\bar{S}(t_f, t)$, defined in (5.169), satisfies the same Riccati differential equation as (5.78) by performing time differentiation of $\bar{S}(t_f, t)$ using (5.167).

6. Consider the problem of finding $u(\cdot) \in \mathcal{U}$ that minimizes

$$J = \int_0^{t_f} \left[\frac{1 + u^2}{1 + x^2} \right]^{1/2} dt$$

subject to $\dot{x} = u$, $x(0) = 0$, $x(t_f) = 0$.

a. Show that the extremal path and control

$$x^o(t) = 0, \quad u^o(t) = 0 \quad \text{for} \quad t \in [0, t_f]$$

satisfies the first-order necessary conditions.

b. By using the second variation, determine if x^o and u^o are locally minimizing.

7. Consider the frequency domain input/output relation

$$y(s) = \frac{1}{s(s+2)}u(s), \tag{5.296}$$

where y = output and u = input.

(a) Obtain a minimal-order state representation for (5.296).

(b) Consider the cost functional

$$J = \int_0^\infty (y^2 + u^2)dt. \tag{5.297}$$

A feedback law of the form

$$u = -kx \tag{5.298}$$

is desired. Determine the constant vector k such that J is minimized.

(c) Calculate the transfer function

$$k^T(sI - A)^{-1}B. \tag{5.299}$$

What is the interpretation of this transfer function? Prove that this transfer function will be the same regardless of the realization used.

8. Solve the optimization problem

$$\min_u J = x^2(t_f) + \int_0^{t_f} u^2(t)dt, \quad u \in \mathcal{U}, \tag{5.300}$$

such that $\dot{x} = u$ and $x(0) = x_0$ $x, u \in \mathbb{R}^1$. Determine u^o as a function of the time t and initial condition x_0. Verify that the Hamiltonian is constant along the path.

9. Derive the first-order necessary conditions of optimality for the following problem:

$$\min_u J(t_0) = \frac{1}{2}x^T(t_f)S(t_f)x(t_f) + \frac{1}{2}\int_{t_0}^{t_f} \left(x^T(t)Q(t)x(t) + u^T(t)R(t)u(t)\right) dt$$

such that

$$\dot{x}(t) = A(t)x(t) + B(t)u(t) + d(t)$$

and $d(t) \in \mathbb{R}^n$ is a *known* disturbance.

10. Solve the following tracking problem using the steady state solution for the Riccati equation

$$\min_u J = \frac{1}{2}p\left[x(t_f) - r(t_f)\right]^2 + \frac{1}{2}\int_0^{t_f} [q(x-r)^2 + u^2]dt \tag{5.301}$$

subject to

$$\dot{x} = ax + bu \quad (x, u \in \mathbb{R}^1) \tag{5.302}$$

and a reference input

$$r(t) = \begin{cases} 0, & t < 0, \\ e^{-t}, & t > 0. \end{cases}$$

Determine the optimum control u^o, the optimum trajectory, and the tracking error. Examine the influence of the weighting parameter q.

Chapter 6
Linear Quadratic Differential Games

6.1 Introduction

In the previous developments, the cost criterion was minimized with respect to the control. In this section a generalization is introduced where there are two controls, one designed to minimize the cost criterion and the other to maximize it. We first consider the LQ problem and develop necessary conditions for optimality. In particular, we show that the optimal solution satisfies a saddle point inequality. That is, if either player does not play his optimal strategy, then the other player gains. Since the problem is linear in the dynamics and quadratic in the cost criterion, we show that this saddle point can be obtained as a perfect square in the adversaries strategies. By following the procedure given in Section 5.4.4, a quadratic function of the state is assumed to be the optimal value of the cost. This function is then used to complete the squares, whereby an explicit form is obtained in a perfect square of the adversaries strategies. Note that in the game problem strategies that are functions of the state as well as time are sought.

The issue of feedback control when the state is not perfectly known is considered next. Our objective is to derive a control synthesis method called H_∞ synthesis and as a parameter goes to zero the H_2 synthesis method is recovered. To this end a disturbance attenuation function is defined as an input–output transfer function representing the ratio of a quadratic norm of the desired system outputs over the quadratic norm of the input disturbances. It is shown that under certain conditions the disturbance attenuation function can be bounded. This is shown by formulating a related differential game problem, the solution to which satisfies the desired bound for the disturbance attenuation problem. The results produce a linear controller based upon a measurement sequence. This controller is then specialized to show an explicit full state feedback and a linear estimator for the state. Relationships with current synthesis algorithms are then made.

6.2 LQ Differential Game with Perfect State Information

The linear dynamic equation is extended to include an additional control vector w as

$$\dot{x}(t) = A(t)x(t) + B(t)u(t) + \Gamma(t)w(t) \quad , \quad x(t_0) = x_0. \tag{6.1}$$

The problem is to find control $u(\cdot) \in \mathcal{U}$ which minimizes, and $w(\cdot) \in \mathcal{W}$ ($\mathcal{U}$ and $\mathcal{W}$ are similar admissible sets defined in Assumption 3.2.2) which maximizes the performance criterion

$$\begin{aligned} J(u(\cdot), w(\cdot); x_0, t_0) \quad = \quad & \frac{1}{2}x^T(t_f)Q_f x(t_f) + \frac{1}{2}\int_{t_0}^{t_f} (x^T(t)Q(t)x(t) \\ & + u^T(t)R(t)u(t) - \theta w^T(t)W^{-1}(t)w(t))dt, \end{aligned} \tag{6.2}$$

where $x(t) \in \mathbb{R}^n$, $u(t) \in \mathbb{R}^m$, $w(t) \in \mathbb{R}^p$, $Q^T(t) = Q(t) \geq 0$, $R^T(t) = R(t) > 0$, $W^T(t) = W(t) > 0$, and $Q_f^T = Q_f$. Note that if the parameter $\theta < 0$, the players $u(\cdot)$ and $w(\cdot)$ are cooperative ($w(\cdot)$ minimizes the cost criterion (6.2)), and we revert back to the results of the last chapter. If $\theta > 0$, then the players $u(\cdot)$ and $w(\cdot)$ are adversarial where $u(\cdot)$ minimizes and $w(\cdot)$ maximizes $J(u(\cdot), w(\cdot); x_0, t_0)$. Note the negative weight in the cost penalizes large excursions in $w(\cdot)$. We consider only $\theta > 0$.

Unlike the earlier minimization problems, a saddle point inequality is sought such that

$$J(u^\circ(\cdot), w(\cdot); x_0, t_0) \le J(u^\circ(\cdot), w^\circ(\cdot); x_0, t_0) \le J(u(\cdot), w^\circ(\cdot); x_0, t_0). \tag{6.3}$$

The functions $(u^\circ(\cdot), w^\circ(\cdot))$ are called saddle point controls or strategies. If either player deviates from this strategy, the other player gains. This is also called a zero sum game, since whatever one player loses, the other gains. For these strategies to be useful, the strategies should be functions of both state and time.

We assume that the saddle point value of the cost is given by the optimal value function $J(u^\circ(\cdot), w^\circ(\cdot); x_0, t_0) \stackrel{\triangle}{=} V(x(t), t) = \frac{1}{2}x^T(t)S_G(t_f, t; Q_f)x(t)$, where $S_G(t_f, t; Q_f)$ will be generated by a Riccati differential equation consistent with the game formulation. Our objective is to determine the form of the Riccati differential equation. This choice for the optimal value functions seems natural, given that the optimal value function for the LQ problem in Chapter 5 are all quadratic functions of the state. Note that since only the symmetric part of $S_G(t_f, t; Q_f)$ contributes to the quadratic form, only the symmetric part is assumed.

We use a procedure suggested in Section 5.4.4 to complete the square of a quadratic form. We add the identity

$$\begin{aligned}
&-\int_{t_0}^{t_f} x^T(t)S_G(t_f, t; Q_f)\dot{x}(t)dt \\
&\quad = \frac{1}{2}\int_{t_0}^{t_f} x^T(t)\dot{S}_G(t_f, t; Q_f)x(t)dt - \frac{1}{2}x^T(t)S_G(t_f, t; Q_f)x(t)\Big|_{t_0}^{t_f}
\end{aligned}$$

to (6.2) as

$$
\begin{aligned}
&\hat{J}(u(\cdot), w(\cdot); S_G(t_f, t; Q_f), x_0, t_0) \\
&= \int_{t_0}^{t_f} \Big[\frac{1}{2}(x^T(t)Q(t)x(t) + u^T(t)R(t)u(t) - \theta w^T(t)W^{-1}(t)w(t)) \\
&\qquad + x^T(t)S_G(t_f, t; Q_f)(A(t)x(t) + B(t)u(t) + \Gamma(t)w(t)) \\
&\qquad + \frac{1}{2}x^T(t)\dot{S}_G(t_f, t; Q_f)x(t) \Big] \, dt \\
&\quad - \frac{1}{2}x^T(t)S_G(t_f, t; Q_f)x(t)\Big|_{t_0}^{t_f} + \frac{1}{2}x^T(t_f)Q_f x(t_f) \\
&= \int_{t_0}^{t_f} \Big[\frac{1}{2}x^T(t)(Q(t) + S_G(t_f, t; Q_f)A(t) + A^T(t)S_G(t_f, t; Q_f) \\
&\qquad + \dot{S}_G(t_f, t; Q_f))x(t) + \frac{1}{2}u^T(t)R(t)u(t) - \frac{1}{2}\theta w^T(t)W^{-1}(t)w(t) \\
&\qquad + x^T(t)S_G(t_f, t; Q_f)(B(t)u(t) + \Gamma(t)w(t)) \Big] \, dt \\
&\quad - \frac{1}{2}x^T(t)S_G(t_f, t; Q_f)x(t)\Big|_{t_0}^{t_f} + \frac{1}{2}x^T(t_f)Q_f x(t_f).
\end{aligned} \tag{6.4}
$$

By choosing $S_G(t_f, t; Q_f)$ to satisfy the matrix Riccati equation

$$
\begin{aligned}
&Q(t) + S_G(t_f, t; Q_f)A(t) + A^T(t)S_G(t_f, t; Q_f) + \dot{S}_G(t_f, t; Q_f) \\
&\quad = S_G(t_f, t; Q_f)(B(t)R^{-1}(t)B^T(t) - \theta^{-1}\Gamma(t)W(t)\Gamma^T(t))S_G(t_f, t; Q_f)\ , \\
&\qquad S_G(t_f, t_f; Q_f) = Q_f,
\end{aligned} \tag{6.5}
$$

$\hat{J}(u(\cdot), w(\cdot); S_G(t_f, t; Q_f), x_0, t_0)$ reduces to squared terms in the strategies of $u(\cdot)$ and $w(\cdot)$ as

$$
\begin{aligned}
&\hat{J}(u(\cdot), w(\cdot); S_G(t_f, t; Q_f), x_0, t_0) \\
&= \frac{1}{2}\int_{t_0}^{t_f} \Big[(u(t) + R^{-1}(t)B^T(t)S_G(t_f, t; Q_f)x(t))^T R(t)(u(t) + R^{-1}(t)B^T(t) \\
&\quad \times S_G(t_f, t; Q_f)x(t)) - (w(t) - \theta^{-1}W(t)\Gamma^T(t)S_G(t_f, t; Q_f)x(t))^T \theta W^{-1} \\
&\quad \times (w(t)\theta^{-1}W(t)\Gamma^T(t)S_G(t_f, t; Q_f)x(t)) \Big] dt + \frac{1}{2}x^T(t_0)S_G(t_f, t_0; Q_f)x(t_0).
\end{aligned} \tag{6.6}
$$

The saddle point inequality (6.3) is satisfied if

$$u^\circ(t) = -R^{-1}(t)B^T(t)S_G(t_f,t;Q_f)x(t), \quad w^\circ(t) = \theta^{-1}W(t)\Gamma^T(t)S_G(t_f,t;Q_f)x(t), \quad (6.7)$$

and the the solution to the Riccati equation (6.5) remains bounded. Since we have completed the square in the cost criterion, Equation (6.6) produces a sufficiency condition for saddle point optimality.

Remark 6.2.1 *The solution to the controller Riccati equation* (6.5) *can have a finite escape time because the matrix* $(B(t)R^{-1}(t)B^T(t) - \theta^{-1}\Gamma(t)W(t)\Gamma^T(t))$ *can be indefinite. Note that the cost criterion* (6.2) *may be driven to large positive values by* $w(\cdot)$*, since the cost criterion is not a concave functional with respect to* $x(t)$ *and* $w(t)$*. Therefore, if* $S_G(t_f,t;Q_f)$ *escapes as* $t \to t_e$*, where* t_e *is the escape time, then for some* $x(t)$ *as* $t \to t_e$*, the cost criterion approaches infinity. There exists a solution* $u^\circ(\cdot) \in \mathcal{U}$ *and* $w^\circ(\cdot) \in \mathcal{W}$ *given in* (6.7) *to the differential game problem if and only if* $S_G(t_f,t;Q_f)$ *exists for all* $t \in [t_0,t_f]$*. This is proved for a more general problem in Theorem* 6.3.1.

6.3 Disturbance Attenuation Problem

We now use the game theoretic results of the last section to develop a controller which is to some degree insensitive to input process and measurement disturbances. Consider the setting in Figure 6.1. The objective is to design a compensator based only on the measurement history, such that the transmission from the disturbances to the performance outputs are limited in some sense. To make these statements more explicit, consider the dynamic system

$$\dot{x}(t) = A(t)x(t) + B(t)u(t) + \Gamma(t)w(t), \quad x(t_0) = x_o, \quad (6.8)$$

$$z(t) = H(t)x(t) + v(t), \quad (6.9)$$

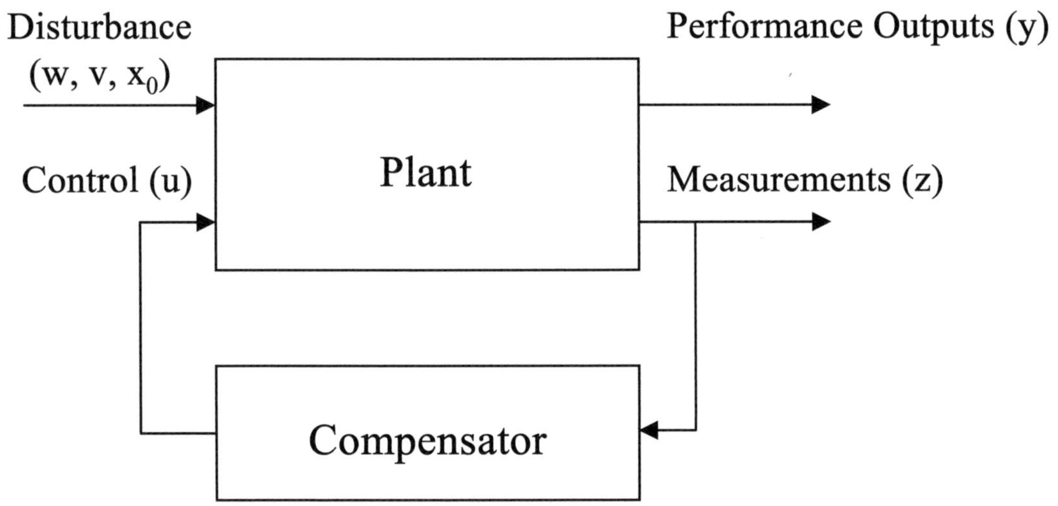

Figure 6.1: Disturbance attenuation block diagram.

where $z(t) \in \mathbb{R}^q$ is the measurement, $w(t) \in \mathbb{R}^m$ is the process disturbance error, $v(t) \in \mathbb{R}^q$ is the measurement disturbance error, and $x_0 \in \mathbb{R}^n$ is an unknown initial condition. The matrices $A(t)$, $B(t)$, $\Gamma(t)$, and $H(t)$ are known functions of time. The performance outputs are measures of desired system performance, such as good tracking error or low actuation inputs to avoid saturation. The general performance measure can be written as

$$y(t) = C(t)x(t) + D(t)u(t), \tag{6.10}$$

where $y(t) \in \mathbb{R}^r$.

A general representation of the input-output relationship between disturbances $\tilde{w}(t) = [w^T(t), v^T(t), x_0^T]^T$ and output performance measure $y(t)$ is the disturbance attenuation function

$$D_a = \frac{\|y(\cdot)\|_2^2}{\|\tilde{w}(\cdot)\|_2^2}, \tag{6.11}$$

where we have extended the norm $\|y(u(\cdot))\|_2^2$ to include a quadratic terminal function as

$$\|y(\cdot)\|_2^2 \triangleq \frac{1}{2}\Bigg[x^T(t_f)Q_f x(t_f) + \int_{t_0}^{t_f}(x^T(t)Q(t)x(t) + u^T(t)R(t)u(t))dt\Bigg], \tag{6.12}$$

where the integrand is

$$\|y(t)\|_2^2 \triangleq y^T(t)y(t) = x^T(t)C^T(t)C(t)x(t) + u^T(t)D^T(t)D(t)u(t),$$
$$C^T(t)C(t) = Q(t),\;\; C^T(t)D(t) = 0,\;\; D^T(t)D(t) = R(t), \tag{6.13}$$

and $\tilde{w}(t) \triangleq [w^T(t), v^T(t), x_0^T]^T$, where, with $V(t) = V^T(t) > 0$ and $P_0 = P_0^T > 0$,

$$\|\tilde{w}(\cdot)\|_2^2 \triangleq \frac{1}{2}\left[\int_{t_0}^{t_f}(w^T(t)W^{-1}(t)w(t) + v^T(t)V^{-1}(t)v(t))dt + x_0^T P_0^{-1}x_0\right]. \tag{6.14}$$

In this formulation, we assume that the cross terms for the disturbances $w(t)$ and $v(t)$ in (6.14) are zero.

The disturbance attenuation problem is to find a controller $u(t) = u(Z_t) \in \mathcal{U}_\ell \subset \mathcal{U}$, where the measurement history is $Z_t \triangleq \{z(s) : 0 \le s \le t\}$, so that the disturbance attenuation problem is bounded as

$$D_a \le \theta, \quad \theta > 0, \tag{6.15}$$

for all admissible processes of $w(t)$ and $v(t)$ and initial condition $x_0 \in \mathbb{R}^n$. $\mathcal{U}$ is the class of admissible controls and $\mathcal{U}_\ell \subset \mathcal{U}$ is the subset of controllers that are linear functions of Z_t. The choice of θ cannot be completely arbitrary. There exists a θ_c where if $\theta \le \theta_c$, the solution to the problem does not exist. As will be shown, the solutions to associated Riccati equations may have a finite escape time.

6.3.1 The Disturbance Attenuation Problem Converted into a Differential Game

This disturbance attenuation problem is converted to a differential game problem with performance index obtained from manipulating Equations (6.11) and (6.15) as

$$J(u(\cdot),\tilde{w}(\cdot);t_0,t_f) = \|y(\cdot)\|_2^2 - \theta\|\tilde{w}(\cdot)\|_2^2. \tag{6.16}$$

For convenience define a process for a function $\hat{w}(\cdot)$ as

$$\hat{w}_a^b(\cdot) \triangleq \{\hat{w}(t) : a \le t \le b\}. \tag{6.17}$$

The differential game is then to find the minimax solution as

$$J^\circ(u^\circ(\cdot),\tilde{w}^\circ(\cdot);t_0,t_f) = \min_{u_{t_0}^{t_f}(\cdot)} \max_{w_{t_0}^{t_f}(\cdot),v_{t_0}^{t_f}(\cdot),x_0} J(u(\cdot),\tilde{w}(\cdot);t_0,t_f). \tag{6.18}$$

We first assume that the min and max operations are interchangeable. It can be shown [40] that the solution has a saddle point, and therefore this interchange of the min and max operations is valid. The saddle point condition is validated in Section 6.3.3. This problem is solved by dividing the problem into a future part, $\tau > t$, and past part, $\tau < t$, and joining them together with a connection condition, where t is the "current" time. Therefore, expand Equation (6.18) as

$$\begin{aligned} &J^\circ(u^\circ(\cdot),\tilde{w}^\circ(\cdot);t_0,t_f) \\ &= \min_{u_{t_0}^{t}(\cdot)} \max_{w_{t_0}^{t}(\cdot),v_{t_0}^{t}(\cdot),x_0} \left[J(u(\cdot),\tilde{w}(\cdot);t_0,t) + \min_{u_t^{t_f}(\cdot)} \max_{w_t^{t_f}(\cdot),v_t^{t_f}(\cdot)} J(u(\cdot),\tilde{w}(\cdot);t,t_f) \right]. \end{aligned} \tag{6.19}$$

Note that for the future time interval no measurements are available. Therefore, maximizing with respect to $v_t^{t_f}(\cdot)$ given the form of the performance index (6.14) and (6.16) produces the worst future process if $v_t^{t_f}(\cdot)$ is given as

$$v(\tau) = 0, \qquad t < \tau \le t_f. \tag{6.20}$$

Therefore, the game problem associated with the future reduces to a game between only $u_t^{t_f}(\cdot)$ and $w_t^{t_f}(\cdot)$. The results given in Section 6.2 are now applicable. In particular, the controller of Equation (6.7) is

$$u^\circ(t) = -R^{-1}(t)B^T(t)S_G(t_f, t; Q_f)x(t), \tag{6.21}$$

where $x(t)$ is *not* known. The objective is to determine $x(t)$ as a function of the measurement history by solving the problem associated with the past. Note that the optimal value function at t is

$$V(x(t), t) = \frac{1}{2}x(t)^T S_G(t_f, t; Q_f)x(t) \tag{6.22}$$

from Equation (6.6), where t, the current time, rather than t_0, is considered the initial time.

6.3.2 Solution to the Differential Game Problem Using the Conditions of the First-Order Variations

Using Equation (6.22) to replace the second term in (6.19), the problem reduces to

$$J^\circ(u^\circ(\cdot), \tilde{w}^\circ(\cdot); t_0, t_f) = \min_{u_{t_0}^t(\cdot)} \max_{w_{t_0}^t(\cdot), v_{t_0}^t(\cdot), x_0} [J(u(\cdot), \tilde{w}(\cdot); t_0, t) + V(x(t), t)]. \tag{6.23}$$

This game problem can be reduced further by noting that $u_{t_0}^t(\cdot)$ has already occurred and therefore, the minimization is meaningless. Second, if the cost criterion is maximized with respect to $w_{t_0}^t$ and x_0, then the resulting state is determined.

Since the state history Z_t is known, then the value of $v_{t_0}^t$ is a result and is eliminated in the cost criterion by using Equation (6.9) as

$$v(\tau) = z(\tau) - H(\tau)x(\tau), \qquad t_0 \le \tau < t. \tag{6.24}$$

The optimization problem now reduces to

$$
\begin{aligned}
&J^\circ(u^\circ(\cdot), \tilde{w}^\circ(\cdot); t_0, t_f) \\
&\quad = \max_{w_{t_0}^t, x_0} \frac{1}{2}\Bigg[\int_{t_0}^{t} (x^T(\tau)Q(\tau)x(\tau) + u^T(\tau)R(\tau)u(\tau))d\tau \\
&\qquad\qquad - \theta \int_{t_0}^{t} (w^T(\tau)W^{-1}(\tau)w(\tau) + (z(\tau) - H(\tau)x(\tau))^T V^{-1}(\tau)(z(\tau) \\
&\qquad\qquad - H(\tau)x(\tau)))d\tau - \theta x_0^T P_0^{-1} x_0 + x(t)^T S_G(t_f, t; Q_f)x(t)\Bigg]
\end{aligned} \tag{6.25}
$$

subject to the dynamic equations

$$
\dot{x}(t) = A(t)x(t) + B(t)u(t) + \Gamma(t)w(t), \quad x(0) = x_0. \tag{6.26}
$$

In a manner similar to that of Section 3.3.1, the dynamics (6.26) are augmented to the performance index (6.25) with the Lagrange multiplier $\lambda(t)$. The augmented performance index is

$$
\begin{aligned}
&\hat{J}^\circ(u^\circ(\cdot), \tilde{w}^\circ(\cdot); t_0, t_f) \\
&\quad = \max_{w_{t_0}^t, x_0} \Bigg[\int_{t_0}^{t} \frac{1}{2}\Big[x^T(\tau)Q(\tau)x(\tau) + u^T(\tau)R(\tau)u(\tau) - \theta(w^T(\tau)W^{-1}(\tau)w(\tau) + (z(\tau) \\
&\qquad\qquad - H(\tau)x(\tau))^T V^{-1}(\tau)(z(\tau) - H(\tau)x(\tau))) + 2\lambda^T(\tau)(A(\tau)x(\tau) \\
&\qquad\qquad + B(\tau)u(\tau) + \Gamma(\tau)w(\tau) - \dot{x}(\tau))\Big]d\tau \\
&\qquad - \theta\frac{1}{2}x_0^T P_0^{-1} x_0 + \frac{1}{2}x^T(t)S_G(t_f, t; Q_f)x(t)\Bigg].
\end{aligned} \tag{6.27}
$$

Integrate by parts to give

$$
\begin{aligned}
&\hat{J}^\circ(u^\circ(\cdot), \tilde{w}^\circ(\cdot); t_0, t_f) \\
&\quad = \max_{w_{t_0}^t, x_0} \Bigg[\int_{t_0}^{t} \frac{1}{2}\Big[x^T(\tau)Q(\tau)x(\tau) + u^T(\tau)R(\tau)u(\tau) - \theta(w^T(\tau)W^{-1}(\tau)w(\tau) + (z(\tau) \\
&\qquad\qquad - H(\tau)x(\tau))^T V^{-1}(\tau)(z(\tau) - H(\tau)x(\tau))) + 2\lambda^T(\tau)(A(\tau)x(\tau) \\
&\qquad\qquad + B(\tau)u(\tau) + \Gamma(\tau)w(\tau)) + 2\dot{\lambda}^T(\tau)x(\tau)\Big]d\tau \\
&\qquad - \lambda^T x\Big|_{t_0}^{t} - \theta\frac{1}{2}x_0^T P_0^{-1} x_0 + \frac{1}{2}x^T(t)S_G(t_f, t; Q_f)x(t)\Bigg].
\end{aligned} \tag{6.28}
$$

By taking the first variation of $\hat{J}^\circ(u^\circ(\cdot), \tilde{w}^\circ(\cdot); t_0, t_f)$ as was done in Section 3.3.1, we obtain

$$\begin{aligned}
\delta\hat{J} &= \hat{J}(u^\circ(\cdot), \tilde{w}(\cdot); t_0, t_f) - \hat{J}^\circ(u^\circ(\cdot), \tilde{w}^\circ(\cdot); t_0, t_f)\\
&= \int_{t_0}^{t} \Big[x^T(\tau)Q(\tau)\delta x(\tau)\\
&\quad - \theta\Big[w^T(\tau)W^{-1}(\tau)\delta w(\tau) - (z(\tau) - H(\tau)x(\tau))^T V^{-1}(\tau)H(\tau)\delta x(\tau)\Big]\\
&\quad + \lambda^T(\tau)A(\tau)\delta x(\tau) + \lambda^T(\tau)\Gamma(\tau)\delta w(\tau) + \dot{\lambda}^T(\tau)\delta x(\tau)\Big] d\tau\\
&\quad - \lambda^T \delta x\Big|_{t_0}^{t} - \theta x_0^T P_0^{-1}\delta x_0 + x^T(t)S_G(t_f, t; Q_f)\delta x(t),
\end{aligned} \tag{6.29}$$

and then the first-order necessary conditions are

$$\dot{\lambda}^T(\tau) + \lambda^T(\tau)A(\tau) + \theta(z(\tau) - H(\tau)x(\tau))^T V^{-1}(\tau)H(\tau) + x^T(\tau)Q(\tau) = 0, \tag{6.30}$$

$$\lambda^T(t_0) - \theta x_0^T P_0^{-1} = 0 \Rightarrow \lambda(t_0) = \theta P_0^{-1}x_0, \tag{6.31}$$

$$-\lambda^T(t) + x^T(t)S_G(t_f, t; Q_f) = 0. \tag{6.32}$$

$$w(\tau) = \theta^{-1}W(\tau)\Gamma^T(\tau)\lambda(\tau), \tag{6.33}$$

where τ is the running variable. The dynamic equations for the Hamiltonian system are

$$\dot{x}(\tau) = A(\tau)x(\tau) + B(\tau)u(\tau) + \theta^{-1}\Gamma(\tau)W(\tau)\Gamma^T(\tau)\lambda(\tau)\ , \quad x(t_0) = x_0, \tag{6.34}$$

$$\begin{aligned}
\dot{\lambda}(\tau) &= -A^T(\tau)\lambda(\tau) - Q(\tau)x(\tau) - H^T(\tau)\theta V^{-1}(\tau)(z(\tau) - H(\tau)x(\tau))\ ,\\
&\quad \lambda(t_0) = \theta P_0^{-1}x_0,
\end{aligned} \tag{6.35}$$

where over the interval $t_0 \le \tau \le t$, $u(\tau)$ and $z(\tau)$ are known processes.

By examination of (6.31), where $\hat{x}(t_0) = 0$, we conjecture and verify in Section 6.3.3 that the following form can be swept forward:

$$\lambda(\tau) = \theta P^{-1}(t_0, \tau; P_0)(x(\tau) - \hat{x}(\tau)) \tag{6.36}$$

or that the solution for $x(\tau)$ satisfies

$$x(\tau) = \hat{x}(\tau) + \theta^{-1} P(t_0, \tau; P_0)\lambda(\tau). \tag{6.37}$$

To determine the differential equations for $\hat{x}(\tau)$ and $P(t_0, \tau; P_0)$, we differentiate (6.37) to get

$$\dot{x}(\tau) = \dot{\hat{x}}(\tau) + \theta^{-1}\dot{P}(t_0, \tau; P_0)\lambda(\tau) + \theta^{-1}P(t_0, \tau; P_0)\dot{\lambda}(\tau). \tag{6.38}$$

Substitute (6.34) and (6.35) into (6.38) we obtain

$$\begin{aligned} & A(\tau)x(\tau) + B(\tau)u(\tau) + \theta^{-1}\Gamma(\tau)W(\tau)\Gamma^T(\tau)\lambda(\tau) \\ & \quad = \dot{\hat{x}}(\tau) + \theta^{-1}\dot{P}(t_0, \tau; P_0)\lambda(\tau) + \theta^{-1}P(t_0, \tau; P_0)(-A^T(\tau)\lambda(\tau) - Q(\tau)x(\tau) \\ & \qquad - \theta H^T(\tau)V^{-1}(\tau)(z(\tau) - H(\tau)x(\tau))). \end{aligned} \tag{6.39}$$

Replace $x(\tau)$ using Equation (6.37) in (6.39). Then

$$\begin{aligned} & A(\tau)\hat{x}(\tau) + \theta^{-1}A(\tau)P(t_0, \tau; P_0)\lambda(\tau) + B(\tau)u(\tau) + \theta^{-1}\Gamma(\tau)W(\tau)\Gamma^T(\tau)\lambda(\tau) \\ & \quad = \dot{\hat{x}}(\tau) + \theta^{-1}\dot{P}(t_0, \tau; P_0)\lambda(\tau) - \theta^{-1}P(t_0, \tau; P_0)A^T(\tau)\lambda(\tau) \\ & \qquad - \theta^{-1}P(t_0, \tau; P_0)Q(\tau)\hat{x}(\tau) - \theta^{-1}P(t_0, \tau; P_0)Q(\tau)\theta^{-1}P(t_0, \tau; P_0)\lambda(\tau) \\ & \qquad - \theta^{-1}P(t_0, \tau; P_0)\theta H^T(\tau)V^{-1}(\tau)(z(\tau) - H(\tau)\hat{x}(\tau)) \\ & \qquad + \theta^{-1}P(t_0, \tau; P_0)\theta H^T(\tau)V^{-1}(\tau)H(\tau)\theta^{-1}P(t_0, \tau; P_0)\lambda(\tau). \end{aligned} \tag{6.40}$$

Rewriting so that all terms multiplying $\lambda(\tau)$ are on one side of the equal sign,

$$\begin{aligned} & A(\tau)\hat{x}(\tau) - \dot{\hat{x}}(\tau) + B(\tau)u(\tau) + \theta^{-1}P(t_0, \tau; P_0)Q(\tau)\hat{x}(\tau) \\ & \quad + \theta^{-1}\theta P(t_0, \tau; P_0)H^T(\tau)V^{-1}(\tau)(z(\tau) - H(\tau)\hat{x}(\tau)) \\ & \quad = -\theta^{-1}A(\tau)P(t_0, \tau; P_0)\lambda(\tau) - \theta^{-1}\Gamma(\tau)W(\tau)\Gamma^T(\tau)\lambda(\tau) + \theta^{-1}\dot{P}(t_0, \tau; P_0)\lambda(\tau) \\ & \qquad - \theta^{-1}P(t_0, \tau; P_0)A^T(\tau)\lambda(\tau) - \theta^{-1}P(t_0, \tau; P_0)Q(\tau)\theta^{-1}P(t_0, \tau; P_0)\lambda(\tau) \\ & \qquad + \theta^{-1}P(t_0, \tau; P_0)\theta H^T(\tau)V^{-1}(\tau)\theta^{-1}H(\tau)P(t_0, \tau; P_0)\lambda(\tau). \end{aligned} \tag{6.41}$$

If we choose $\hat{x}(\tau)$ to satisfy

$$\begin{aligned}\dot{\hat{x}}(\tau) &= A(\tau)\hat{x}(\tau) + B(\tau)u(\tau) + \theta^{-1}P(t_0,\tau;P_0)Q(\tau)\hat{x}(\tau)\\ &\quad + P(t_0,\tau;P_0)H^T(\tau)V^{-1}(\tau)(z(\tau) - H(\tau)\hat{x}(\tau)), \quad \hat{x}(t_0) = 0, \end{aligned} \tag{6.42}$$

and $P(t_0, \tau; P_0)$ to satisfy

$$\begin{aligned}&- A(\tau)P(t_0,\tau;P_0) - \Gamma(\tau)W(\tau)\Gamma^T(\tau) + \dot{P}(t_0,\tau;P_0) - P(t_0,\tau;P_0)A^T(\tau)\\ &- \theta^{-1}P(t_0,\tau;P_0)Q(\tau)P(t_0,\tau;P_0) + P(t_0,\tau;P_0)H^T(\tau)V^{-1}(\tau)H(\tau)P(t_0,\tau;P_0) = 0\end{aligned} \tag{6.43}$$

or

$$\begin{aligned}\dot{P}(t_0,\tau;P_0) &= A(\tau)P(t_0,\tau;P_0) + P(t_0,\tau;P_0)A^T(\tau) + \Gamma(\tau)W(\tau)\Gamma^T(\tau)\\ &\quad - P(t_0,\tau;P_0)(H^T(\tau)V^{-1}(\tau)H(\tau) - \theta^{-1}Q(\tau))P(t_0,\tau;P_0)\ ,\\ &\quad P(t_0) = P_0,\end{aligned} \tag{6.44}$$

then (6.38) becomes an identity.

Remark 6.3.1 *The solution to the estimation Riccati equation* (6.44) *may have a finite escape time because the matrix* $(H^T(\tau)V^{-1}(\tau)H(\tau) - \theta^{-1}Q(\tau))$ *can be indefinite. Some additional properties of the Riccati differential equation of* (6.5) *and* (6.44) *are given in* [40].

At the current time t from Equation (6.32), $\lambda(t) = S_G(t_f, t; Q_f)x(t)$. Therefore, the worst-case state $x^\circ(t)$ is equivalent to $x(t)$ when using Equation (6.37), so that

$$x^\circ(t) = \hat{x}(t) + \theta^{-1}P(t_0,t;P_0)S_G(t_f,t;Q_f)x^\circ(t). \tag{6.45}$$

This is explicitly shown in Section 6.3.3. In factored form from (6.45), the estimate $\hat{x}(t)$ and the worst state $x^\circ(t)$ are related as

$$\begin{aligned}\hat{x}(t) &= (I - \theta^{-1}P(t_0,t;P_0)S_G(t_f,t;Q_f))x^\circ(t),\\ x^\circ(t) &= (I - \theta^{-1}P(t_0,t;P_0)S_G(t_f,t;Q_f))^{-1}\hat{x}(t).\end{aligned} \tag{6.46}$$

The state $x^\circ(t)$ in Equation (6.46) is the saddle value of the state that is used in the controller. The optimal controller $u(t) = u(Z_t)$ is now written as

$$u^\circ(t) = -R^{-1}(t)B^T(t)S_G(t_f,t;Q_f)(I-\theta^{-1}P(t_0,t;P_0)S_G(t_f,t;Q_f))^{-1}\hat{x}(t), \quad (6.47)$$

where $S_G(t_f,t;Q_f)$ is determined by integrating (6.5) backward from t_f and $P(t_0,t;P_0)$ is determined by integrating (6.44) forward from t_0.

Note 6.3.1 $\hat{x}(t)$ *summaries the measurement history in an n-vector.*

It should be noted that if all adversaries play their saddle point strategy, then

$$x^\circ(t) = 0, \; v^\circ(t) = 0, \; \hat{x}(t) = 0, \; u^\circ(t) = 0, \; w^\circ(t) = 0, \quad (6.48)$$

which implies that

$$J(u^\circ(\cdot), \tilde{w}(\cdot); t_0, t_f) \le J^\circ(u^\circ(\cdot), \tilde{w}^\circ(\cdot); t_0, t_f) = 0. \quad (6.49)$$

From Equation (6.49) we see that Equation (6.15) is satisfied and the game solution also provides the solution to the disturbance attenuation problem.

Remark 6.3.2 *Note that when* $\theta \to \infty$ *the disturbance attenuation or* H_∞ *controller given by* (6.47) *reduce to the* H_2 *controller*[10]

$$u^\circ(t) = -R^{-1}(t)B^T(t)S_G(t_f,t;Q_f)\hat{x}(t), \quad (6.50)$$

where the controller gains are obtained from the Riccati equation (6.5) *with* $\theta = \infty$ *and is identical to the Riccati equation in Theorem* 5.6.2. *The state is reconstructed from the filter with* $\hat{x} = x^\circ$

$$\dot{\hat{x}}(t) = A(t)\hat{x}(t) + B(t)u(t) + P(t_0,t;P_0)H^T(t), V^{-1}(t)\left(z(t) - H(t)\hat{x}(t)\right), \quad (6.51)$$

[10]In the stochastic setting this is known as the linear-quadratic-Gaussian (LQG) controller.

where the filter gains are determined from (6.44) *with* $\theta = \infty$. *The properties of this Riccati equation are again similar to those given in Theorem* 5.6.2.

6.3.3 Necessary and Sufficient Conditions for the Optimality of the Disturbance Attenuation Controller

In this section are given necessary and sufficient conditions that guarantee that the controller (6.47) satisfies $J(u^\circ(\cdot), \tilde{w}(\cdot); t_0, t_f) \le 0$, i.e., that $J(u^\circ(\cdot), \tilde{w}(\cdot); t_0, t_f)$ is concave with respect to $\tilde{w}(\cdot)$. To do this, the cost criterion is written as the sum of two quadratic forms evaluated at the current time t. One term is the optimal value function $x^T(t)S(t_f, t; Q_f)x(t)/2$, where the cost criterion is swept backward from the terminal time to the current time t. For the second term, we verify that the optimal value function which sweeps the initial boundary function $-\theta x_0^T P_0^{-1} x_0/2$ forward is

$$\tilde{V}(e(t), t) = \frac{\theta}{2} e^T(t) P^{-1}(t_0, t; P_0) e(t), \tag{6.52}$$

where $e(t) = x(t) - \hat{x}(t)$ and $\tilde{V}_x(e(t), t) = \lambda(t) = \theta P^{-1}(t_0, t; P_0)e(t)$ as given in (6.36). The dynamic equation for $e(t)$ is found by subtracting (6.42) from (6.26) as

$$\begin{aligned} \dot{e}(t) &= A(t)e(t) + \Gamma(t)w(t) - \theta^{-1}P(t_0, t; P_0)Q(t)\hat{x}(t) \\ &\quad - P(t_0, t; P_0)H^T(t)V^{-1}(t)(z(t) - H(t)\hat{x}(t)). \end{aligned} \tag{6.53}$$

As in Section 6.2 we complete the squares of the cost criterion (6.25) by adding the identically zero quantity

$$\begin{aligned} 0 &= \int_{t_0}^{t} \left[\theta e^T(\tau) P^{-1}(t_0, \tau; P_0)\dot{e}(\tau) + \frac{\theta}{2} e^T(\tau)\dot{P}^{-1}(t_0, \tau; P_0)e(\tau) \right] d\tau \\ &\quad - \frac{\theta}{2} e^T(\tau) P^{-1}(t_0, \tau; P_0) e(\tau) \Big|_{t_0}^{t} \end{aligned} \tag{6.54}$$

to (6.25) as

$$
\begin{aligned}
&J(u^\circ(\cdot), \tilde{w}^\circ(\cdot), x(t); t_0, t_f) \\
&= \max_{w_{t_0}^t, x_0} \frac{1}{2}\bigg[\int_{t_0}^t (x^T(\tau)Q(\tau)x(\tau) + u^T(\tau)R(\tau)u(\tau)) - \theta(w^T(\tau)W^{-1}(\tau)w(\tau) \\
&\quad + (z(\tau) - H(\tau)x(\tau))^T V^{-1}(\tau)(z(\tau) - H(\tau)x(\tau)) \\
&\quad - 2e^T(\tau)P^{-1}(t_0,\tau;P_0)\dot{e}(\tau) - e^T(\tau)\dot{P}^{-1}(t_0,\tau;P_0)e(\tau))d\tau \\
&\quad - \theta x_0^T P_0^{-1} x_0 - \theta e^T(\tau)P^{-1}(t_0,\tau;P_0)e(\tau)\Big|_{t_0}^{t} + x(t)^T S_G(t_f, t; Q_f)x(t)\bigg],
\end{aligned}
\tag{6.55}
$$

where the terms evaluated at $t = t_0$ cancel in (6.55). The state $x(t)$ is still free and unspecified. In (6.46) the worst-case state is captured by the boundary conditions. Here, we will maximize $J(u^\circ(\cdot), \tilde{w}^\circ(\cdot), x(t); t_0, t_f)$ with respect to $x(t)$. Note that $e(t_0) = x_0$, since we assumed $\hat{x}(t_0) = 0$ only for convenience. Furthermore, the control in the quadratic term is a given process over the interval $[t_0, t_f)$. From (6.44) the RDE for $P^{-1}(t_0, t; P_0)$ is determined as

$$
\begin{aligned}
\dot{P}^{-1}(t_0,t;P_0) &= -P^{-1}(t_0,t;P_0)A(t) - A^T(t)P^{-1}(t_0,t;P_0) \\
&\quad - P^{-1}(t_0,t;P_0)\Gamma(t)W(t)\Gamma^T(t)P^{-1}(t_0,t;P_0) \\
&\quad + (H^T(t)V^{-1}(t)H(t) - \theta^{-1}Q(t)), \qquad P^{-1}(t_0) = P_0^{-1}.
\end{aligned}
\tag{6.56}
$$

First, (6.53) is substituted into (6.55) to obtain

$$
\begin{aligned}
&J(u^\circ(\cdot), \tilde{w}^\circ(\cdot), x(t); t_0, t_f) \\
&= \max_{w_{t_0}^t, x_0} \frac{1}{2}\bigg[\int_{t_0}^t (x^T(\tau)Q(\tau)x(\tau) + u^T(\tau)R(\tau)u(\tau) - \theta w^T(\tau)W^{-1}(\tau)w(\tau) \\
&\quad - \theta(z(\tau) - H(\tau)x(\tau))^T V^{-1}(\tau)(z(\tau) - H(\tau)x(\tau)) \\
&\quad + 2\theta e^T(\tau)P^{-1}(t_0,\tau;P_0)(A(\tau)e(\tau) + \Gamma(\tau)w(\tau) \\
&\quad - \theta^{-1}P(t_0,(\tau);P_0)Q(\tau)\hat{x}(\tau) \\
&\quad - P(t_0,(\tau);P_0)H^T(\tau)V^{-1}(\tau)(z(\tau) - H(\tau)\hat{x}(\tau))
\end{aligned}
$$

$$+\theta e^T(\tau)\dot{P}^{-1}(t_0,\tau;P_0)e(\tau))d\tau - \theta e^T(t)P^{-1}(t_0,t;P_0)e(t)$$
$$+x(t)^T S_G(t_f,t;Q_f)x(t)\Big], \tag{6.57}$$

By adding and subtracting the term $\theta e^T(\tau)P^{-1}(t_0,\tau;P_0)\Gamma(\tau)W(\tau)\Gamma^T(\tau)P^{-1}(t_0,\tau;P_0)e(\tau)$ and substituting in $x(\tau)=\hat{x}(\tau)+e(\tau)$, (6.57) becomes

$$\begin{aligned}
&J(u^\circ(\cdot),\tilde{w}^\circ(\cdot),x(t);t_0,t_f)\\
&= \max_{w_{t_0}^t, x_0} \frac{1}{2}\Big[\int_{t_0}^t (e^T(\tau)Q(\tau)e(\tau)+\hat{x}^T(\tau)Q(\tau)\hat{x}(\tau)+u^T(\tau)R(\tau)u(\tau)-\theta(z(\tau)\\
&\quad - H(\tau)\hat{x}(\tau))^T V^{-1}(\tau)(z(\tau)-H(\tau)\hat{x}(\tau)) - \theta(w^T(\tau)\\
&\quad - e^T(\tau)P^{-1}(t_0,\tau;P_0)\Gamma(\tau)W(\tau))W^{-1}(\tau)(w(\tau)\\
&\quad - W(\tau)\Gamma^T(\tau)P^{-1}(t_0,\tau;P_0)e(\tau))\\
&\quad + \theta e^T(\tau)P^{-1}(t_0,\tau;P_0)\Gamma(\tau)W(\tau)\Gamma^T(\tau)P^{-1}(t_0,\tau;P_0)e(\tau)\\
&\quad + 2\theta e^T(\tau)P^{-1}(t_0,\tau;P_0)A(\tau)e(\tau) - \theta e^T(\tau)H^T(\tau)V^{-1}(\tau)H(\tau)e(\tau)\\
&\quad + \theta e^T(\tau)\dot{P}^{-1}(t_0,\tau;P_0)e(\tau))d\tau - \theta e^T(t)P^{-1}(t_0,t;P_0)e(t)\\
&\quad + x(t)^T S_G(t_f,t;Q_f)x(t)\Big].
\end{aligned} \tag{6.58}$$

Using (6.56) in (6.58) and maximizing with respect to $w_{t_0}^t$, the optimal cost criterion reduces to

$$\begin{aligned}
&J(u^\circ(\cdot),\tilde{w}^\circ(\cdot),x(t);t_0,t_f)\\
&= \frac{1}{2}\left[I(Z_t) - \theta e^T(t)P^{-1}(t_0,t;P_0)e(t) + x^T(t)S_G(t_f,t;Q_f)x(t)\right],
\end{aligned} \tag{6.59}$$

where

$$\begin{aligned}
I(Z_t) &= \int_{t_0}^t \big[\hat{x}^T(\tau)Q(\tau)\hat{x}(\tau)+u^T(\tau)R(\tau)u(\tau)\\
&\quad - \theta(z(\tau)-H(\tau)\hat{x}(\tau))^T V^{-1}(\tau)(z(\tau)-H(\tau)\hat{x}(\tau))\big]\,d\tau,
\end{aligned} \tag{6.60}$$

the maximizing $w(\tau)$ is

$$w^\circ(\tau) = W(\tau)\Gamma^T(\tau)P^{-1}(t_0,\tau;P_0)e(\tau), \tag{6.61}$$

and maximizing (6.60) with respect to $v(\tau)$ using (6.9) gives

$$v^\circ(\tau) = -H(\tau)e(\tau). \tag{6.62}$$

Since the terms under the integral are functions only of the given measurement process, they are known functions over the past time interval. Therefore, the determination of the worst-case state is found by maximizing over the last two terms in $J(u^\circ(\cdot), \tilde{w}^\circ(\cdot), x(t); t_0, t_f)$ of (6.59). Thus, $J_{x(t)}(u^\circ(\cdot), \tilde{w}^\circ(\cdot), x(t); t_0, t_f) = 0$ gives (6.46) and the second variation condition for a maximum gives

$$P^{-1}(t_0, t; P_0) - \theta^{-1} S_G(t_f, t; Q_f) > 0. \tag{6.63}$$

This inequality is known as the **spectral radius condition**. Note that

$$J(u^\circ(\cdot), \tilde{w}^\circ(\cdot), x^\circ(t); t_0, t_f) = J^\circ(u^\circ(\cdot), \tilde{w}^\circ(\cdot); t_0, t_f).$$

We now show that a necessary and sufficient condition for $J(u^\circ(\cdot), \tilde{w}(\cdot), x(t); t_0, t_f)$ to be concave with respect to $\tilde{w}(\cdot), x(t)$ is that the following assumption be satisfied.

Assumption 6.3.1

1. *There exists a solution $P(t_0, t; P_0)$ to the Riccati differential equation* (6.44) *over the interval* $[t_0, t_f]$.

2. *There exists a solution $S_G(t_f, t; Q_f)$ to the Riccati differential equation* (6.5) *over the interval* $[t_0, t_f]$.

3. $P^{-1}(t_0, t; P_0) - \theta^{-1} S_G(t_f, t; Q_f) > 0$ *over the interval* $[t_0, t_f]$.

Remark 6.3.3 *In* [40] *some properties of this class of Riccati differential equation are presented. For example, if $Q_f \geq 0$ (> 0), then $S_G(t_f, t; Q_f) \geq 0$ (> 0) for $t_f \geq t \geq t_0$. If $S_G(t_f, t; Q_f)$ has an escape time in the interval $t_f \geq t \geq t_0$, then some eigenvalues of $S_G(t_f, t; Q_f)$ must go off to positive infinity.*

Theorem 6.3.1 *There exists a solution* $u^\circ(Z_t) \in \mathcal{U}_\ell$ *to the finite-time disturbance attenuation problem if and only if Assumption* 6.3.1 *holds. If Assumption* 6.3.1 *holds,* $u^\circ(t) = -R^{-1}(t)B^T(t)S_G(t_f, t; Q_f)x^\circ(t)$.

Proof: *Sufficiency:* Suppose that Assumption 6.3.1 holds. For the strategies $u^\circ(t) = -R^{-1}(t)B^T(t)S_G(t_f, t; Q_f)x^\circ(t)$, $w^\circ(t) = \theta^{-1}W(t)\Gamma^T(t)S_G(t_f, t; Q_f)x^\circ(t)$, and $v^\circ(t) = -H(t)e(t)$, Equation (6.49) holds. Furthermore, using (6.45) and (6.46), $\hat{x}(t) = 0$, where the strategies are used in (6.42), as well as $e(t) = 0$, where the strategies are used in (6.53). Therefore,

$$J(u^\circ(\cdot), \tilde{w}^\circ(\cdot), x^\circ(t_f); t_0, t_f) = \frac{1}{2}\left[I(Z_{t_f}) - \theta e^T(t_f)P^{-1}(t_0, t_f; P_0)e(t_f)\right] = 0, \quad (6.64)$$

where ($u^\circ(t) = 0$, $\tilde{w}^\circ(t) = 0$) and for any other strategy, where $(\tilde{w}(t)) \neq 0$, $J(u^\circ(\cdot), \tilde{w}(\cdot), x(t_f); t_0, t_f) < 0$ and thus the cost is strictly concave with respect to $\tilde{w}(\cdot)$.

Necessity: Suppose 1 is violated, but 2 and 3 are not, and $P(t_0, t; P_0)$ has an escape time $t_e \in [t_0, t_f]$. For $t_s < t_e$ we choose the strategies for $\tilde{w}(t)$ as $w^\circ(t) = W(t)\Gamma^T(t)P^{-1}(t_0, t; P_0)e(t)$, $v^\circ(t) = -H(t)e(t)$ for $t_0 \leq t < t_s$ and, using (6.32) and (6.33), $w^\circ(t) = \theta^{-1}W(t)\Gamma^T(t)S_G(t_f, t; Q_f)x(t)$, $v^\circ(t) = 0$ for $t_s \leq t \leq t_f$. Therefore, $\hat{x}(t) = 0$ and $u^\circ(t) = 0$ over $t \in [t_0, t_s]$. Furthermore, we assume some $x(t_0) \neq 0$ such that $e(t_s) = x(t_s) \neq 0$ and coincides with the eigenvector associated with the largest eigenvalue of $P(t_0, t_s; P_0)$. Then, as $t_s \to t_e$, $e^T(t_s)P^{-1}(t_0, t_s; P_0)e(t_s) \to 0$ and the cost criterion $J(u^\circ(\cdot), \tilde{w}(\cdot), x(t_s); t_0, t_f) = \frac{1}{2}x^T(t_s)S_G(t_f, t_s; Q_f)x(t_s) > 0$, which is a contradiction to the optimality of $u^\circ(\cdot)$. Note that $I(Z_{t_s}) = 0$.

Suppose 2 is violated, but 1 and 3 are not, and $S_G(t_f, t; Q_f)$ has an escape time $t_e \in [t_0, t_f]$. For $t_s > t_e$ we choose the strategies of $\tilde{w}(t)$ as $w(t) = 0$,

$v^\circ(t) = -H(t)e(t)$ for $t_0 \le t \le t_s$ and $w^\circ(t) = \theta^{-1}W(t)\Gamma^T(t)S_G(t_f,t;Q_f)x(t)$, $v^\circ(t) = 0$ for $t_s \le t \le t_f$. Furthermore, we assume some $x(t_0) \neq 0$ such that $e(t_s) = x(t_s) \neq 0$ and coincides with the eigenvector associated with the largest eigenvalue of $S_G(t_f, t_s; Q_f)$. Then, as $t_s \to t_e$,

$$\begin{aligned} J(u^\circ(\cdot), \tilde{w}(\cdot), x(t); t_0, t_f) &= \frac{1}{2}\left[I(Z_{t_s}) - \theta e^T(t_{t_s})P^{-1}(t_0, t_{t_s}; P_0)e(t_{t_s})\right. \\ &\quad \left. + \; x^T(t_{t_s})S_G(t_f, t_{t_s}; Q_f)x(t_{t_s})\right] > 0, \end{aligned} \tag{6.65}$$

since as $t_s \to t_e$, $S_G(t_f, t_{t_s}; Q_f)$ goes to positive infinity and the third term dominates and produces a contradiction to the optimality of $u^\circ(\cdot)$. Note that the above is true for any finite control process $u(\cdot)$ over $[t_0, t_s]$ as long as $x(t_s) \neq 0$.

Suppose 3 is violated at $t = t_s$, where $t_s \in [t_0, t_f]$, but 1 and 2 are not. Choose the strategies of $\tilde{w}(t)$ as $w^\circ(t) = W(t)\Gamma^T(t)P^{-1}(t_0, t; P_0)e(t)$, $v^\circ(t) = -H(t)e(t)$ for $t_0 \le t \le t_s$ and $w^\circ(t) = \theta^{-1}W(t)\Gamma^T(t)S_G(t_f, t; Q_f)x(t)$, $v^\circ(t) = 0$ for $t_s \le t \le t_f$. Furthermore, choose $x(t_0) \neq 0$ so that $e(t_s) = x(t_s) \neq 0$ so that $e(t_s) = x(t_s)$ is an eigenvector of a negative eigenvalue of $P^{-1}(t_0, t_s; P_0) - \theta^{-1}S_G(t_f, t_s; Q_f)$ so that $e^T(t_s)(S_G(t_f, t_s; Q_f) - \theta P^{-1}(t_0, t_s; P_0))e(t_s) > 0$. This is again a contradiction to the optimality of $u^\circ(\cdot)$. Note that $I(Z_{t_s}) = 0$. ∎

6.3.4 Time-Invariant Disturbance Attenuation Estimator Transformed into the H_∞ Estimator

For convenience, we consider the infinite-time, time-invariant problem. We assume that $P(t_0, t; P_0)$ has converged to a steady state value denoted as P and $S_G(t_f, t; Q_f)$ has converged to a steady state value denoted as S. We first make a transformation of our estimate $\hat{x}(t)$ to a new estimate $x^\circ(t)$, which is essentially the worst-case state estimate as

$$x^\circ(t) = \left[I - \theta^{-1}PS\right]^{-1}\hat{x}(t) = L^{-1}\hat{x}(t), \tag{6.66}$$

where the estimator propagation is written as a standard differential equation as

$$\dot{\hat{x}}(t) = \left(A + \theta^{-1}PQ\right)\hat{x}(t) + Bu + PH^TV^{-1}\left(z(t) - H\hat{x}(t)\right), \tag{6.67}$$

where we are assuming that all the coefficients are time invariant and the matrices P and S are determined from the algebraic Riccati equations (ARE) as

$$0 = AP + PA^T - P(H^TV^{-1}H - \theta^{-1}Q)P + \Gamma W \Gamma^T, \tag{6.68}$$

$$0 = A^TS + SA + Q - S\left(BR^{-1}B^T - \theta^{-1}\Gamma W \Gamma^T\right)S. \tag{6.69}$$

Substitution of the transformation (6.66) into the estimator (6.67) gives

$$L^{-1}\dot{\hat{x}}(t) = \dot{x}^{\circ}(t) = L^{-1}\left(A + \theta^{-1}PQ\right)Lx^{\circ}(t) + L^{-1}Bu + L^{-1}PH^TV^{-1}\left(z(t) - HLx^{\circ}(t)\right). \tag{6.70}$$

Remark 6.3.4 *The convergence of the Riccati differential equation to an ARE is shown in* [40] *and follows similar notions given in Theorem* 5.7.1 *of Section* 5.7 *for the convergence to the ARE for the linear-quadratic problem. Also, note that the solutions to ARE of* (6.68) *and* (6.69) *can have more than one positive definite solution. However, the minimal positive definite solution (M) to the ARE captures the stable eigenvalues. The minimal positive definite solution M means that for any other solution S, $S - M \geq 0$.*

The elements of the transformation L^{-1} can be manipulated into the following forms which are useful for deriving the dynamic equation for $x^{\circ}(t)$:

$$\begin{aligned} E &= S\left[I - \theta^{-1}PS\right]^{-1} = \left[(I - \theta^{-1}PS)S^{-1}\right]^{-1} = \left[S^{-1} - \theta^{-1}P\right]^{-1} \\ &= \left[S^{-1}(I - \theta^{-1}SP)\right]^{-1} = \left[I - \theta^{-1}SP\right]^{-1}S. \end{aligned} \tag{6.71}$$

Furthermore, from (6.71)

$$S^{-1}E = \left[I + \theta^{-1}PE\right] = \left[I - \theta^{-1}PS\right]^{-1} = \left[P^{-1} - \theta^{-1}S\right]^{-1}P^{-1} = L^{-1}. \tag{6.72}$$

Substitution of the transformations of L and L^{-1} from (6.66) and (6.72) into (6.70) gives

$$\begin{aligned}
\dot{x}^{\circ}(t) &= \left[I+\theta^{-1}PE\right]\left(A+\theta^{-1}PQ\right)\left[I-\theta^{-1}PS\right]x^{\circ}(t)+\left[I+\theta^{-1}PE\right]Bu \\
&\quad +MH^TV^{-1}\left(z(t)-H\left[I-\theta^{-1}PS\right]x^{\circ}(t)\right) \\
&= \left[I+\theta^{-1}PE\right]\left(A+\theta^{-1}PQ\right)\left[I-\theta^{-1}PS\right]x^{\circ}(t)+\left[I+\theta^{-1}PE\right]Bu \\
&\quad +MH^TV^{-1}\left(z(t)-Hx^{\circ}(t)\right)-\left[I+\theta^{-1}PE\right]\theta PH^TV^{-1}PSx^{\circ}(t) \\
&= \left[I+\theta^{-1}PE\right]\left(A+\theta^{-1}PQ\right)x^{\circ}(t)+\left[I+\theta^{-1}PE\right]Bu \\
&\quad +MH^TV^{-1}\left(z(t)-Hx^{\circ}(t)\right) \\
&\quad +\left[I+\theta^{-1}PE\right]\left[\left(A+\theta^{-1}PQ\right)\theta P+\theta^{-1}PH^TV^{-1}P\right]Sx^{\circ}(t) \\
&= \left[I+\theta^{-1}PE\right]\left(A+\theta^{-1}PQ\right)x^{\circ}(t)+\left[I+\theta^{-1}PE\right]Bu \\
&\quad +MH^TV^{-1}\left(z(t)-Hx^{\circ}(t)\right)+\theta^{-1}\left[I+\theta^{-1}PE\right]\left[PA^T+\Gamma W\Gamma^T\right]Sx^{\circ}(t),
\end{aligned} \tag{6.73}$$

where $M = L^{-1}P$ and the last line results from using (6.68) in previous equality. To continue to reduce this equation, substitute in the optimal controller $u^{\circ} = -R^{-1}B^TSx^{\circ}(t)$ into (6.73) Then, (6.73) becomes

$$\begin{aligned}
\dot{x}^{\circ}(t) &= Ax^{\circ}(t)-BR^{-1}B^TSx^{\circ}(t)+\theta^{-1}\Gamma W\Gamma^TSx^{\circ}(t)+MH^TV^{-1}\left(z(t)-Hx^{\circ}(t)\right) \\
&\quad +\theta^{-1}\left[I+\theta^{-1}PE\right]PQx^{\circ}(t)+\theta^{-1}\left[I+\theta^{-1}PE\right]PA^TSx^{\circ}(t) \\
&\quad -\theta^{-1}PE\left[-A+BR^{-1}B^TS-\theta^{-1}\Gamma W\Gamma^TS\right]x^{\circ}(t).
\end{aligned} \tag{6.74}$$

Noting that

$$\begin{aligned}
PE &= P\left[I-\theta^{-1}SP\right]^{-1}S=\left[P^{-1}-\theta^{-1}S\right]^{-1}S, \\
\left[I+\theta^{-1}PE\right] &= \left[P^{-1}-\theta^{-1}S\right]^{-1}P^{-1}.
\end{aligned} \tag{6.75}$$

Substituting (6.75) into (6.74) and using (6.69), the estimator in terms of $x^\circ(t)$ becomes

$$\dot{x}^\circ(t) = Ax^\circ(t) - BR^{-1}B^TSx^\circ(t) + \theta^{-1}\Gamma W\Gamma^TSx^\circ(t) + MH^TV^{-1}\left(z(t) - Hx^\circ(t)\right). \tag{6.76}$$

The appearance of the term $w = +\theta^{-1}\Gamma W\Gamma^TSx^\circ(t)$ is the optimal strategy of the process noise and explicitly is included in the estimator. This estimator equation is the same if the system matrices are time varying and is equivalent to that given in [18] for their time-invariant problem. The dynamic equation for the matrix $M(t)$ in the filter gain can be obtained by differentiating $M(t)$ as

$$\begin{aligned} M(t) &= L^{-1}(t)P(t_0,t;P_0) = \left[P^{-1}(t_0,t;P_0) - \theta^{-1}S_G(t_f,t;Q_f)\right]^{-1} \\ &\Rightarrow \dot{M}(t) = M(t)\left[\dot{P}^{-1}(t_0,t;P_0) - \theta^{-1}\dot{S}_G(t_f,t;Q_f)\right]M(t). \end{aligned} \tag{6.77}$$

Substitution of (6.5) and (6.44) into (6.77) produces the Riccati equation

$$\begin{aligned} \dot{M}(t) &= M(t)\left(A - \theta\Gamma W\Gamma^TS_G(t_f,t;Q_f)\right)^T + \left(A - \theta\Gamma W\Gamma^TS_G(t_f,t;Q_f)\right)M(t) \\ &\quad - M(t)\left(H^TV^{-1}H - \theta^{-1}S_G(t_f,t;Q_f)BR^{-1}B^TS_G(t_f,t;Q_f)\right)M(t) \\ &\quad + \Gamma W\Gamma^T, \qquad M(t_0) = \left(I - \theta^{-1}P_0S_G(t_f,t_0;Q_f)\right)^{-1}P_0. \end{aligned} \tag{6.78}$$

For the infinite-time, time-invariant system, $\dot{M} = 0$ and (6.78) becomes an ARE.

Relating this back to the disturbance attenuation controller in the previous section (6.5), (6.42), (6.44), (6.47), the H_∞ form of the disturbance attenuation controller is

$$u^\circ(t) = -R^{-1}B^TSx^\circ(t), \tag{6.79}$$

where $x^\circ(t)$ is given by (6.76), in which M is given by (6.77) and controller and filter gains require the smallest positive definite (see [40]) solutions P and S to the AREs (6.68) and (6.69).

Remark 6.3.5 *Note that when $\theta \to 0$ the H_∞ controller given by* (6.47) *or* (6.79) *reduce to the H_2 controller*[11]

$$u(t) = -R^{-1}B^T S\hat{x}(t), \tag{6.80}$$

where the controller gains are obtained from the ARE (6.69) *with $\theta = 0$ and is identical to the ARE in Theorem* 5.7.2. *The state is reconstructed from the filter with $\hat{x}(t) = x^\circ(t)$:*

$$\dot{\hat{x}}(t) = A\hat{x}(t) + Bu(t) + PH^T V^{-1}\left(z(t) - H\hat{x}(t)\right), \tag{6.81}$$

where the filter gains are determined from (6.68) *with $\theta = 0$. The properties of this new ARE are similar to those given in Theorem* 5.7.2.

6.3.5 H_∞ Measure and H_∞ Robustness Bound

First, we show that the L_2 norms on the input–output of a system induce the H_∞ norm[12] on the resulting transfer matrix. Consider Figure 6.2, where the disturbance input, d, is a square integrable, i.e., L^2 function. We are interested in the conditions on $\mathbf{G}$ that will make the output performance measure, y, square integrable as well. Because of Parseval's Theorem, a square integrable y is isomorphic (i.e., equivalent) to a square integrable transfer function, $\mathbf{Y}(s)$:

$$\|y\|_2^2 = \int_{-\infty}^{\infty} y(\tau)^2 d\tau = \sup_{\alpha>0} \frac{1}{2\pi}\int_{-\infty}^{\infty} \|\mathbf{Y}(\alpha + j\omega)\|^2 d\omega. \tag{6.82}$$

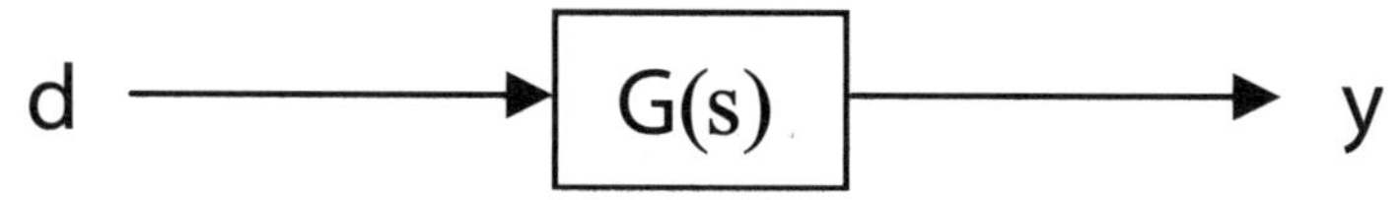

Figure 6.2: Transfer function of square integrable signals.

[11] In the stochastic setting it is known as the linear-quadratic-Gaussian controller.
[12] The "L" in L^2, by the way, stands for "Lebesgue."

We can use the properties of norms and vector spaces to derive our condition on $\mathbf{G}$:

$$
\begin{aligned}
\|y\|_2^2 &= \sup_{\alpha>0} \frac{1}{2\pi}\int_{-\infty}^{\infty} \|\mathbf{G}(\alpha+j\omega)d(\alpha+j\omega)\|^2 d\omega & (6.83)\\
&\leq \sup_{\alpha>0} \frac{1}{2\pi}\int_{-\infty}^{\infty} \|\mathbf{G}(\alpha+j\omega)\|^2\|d(\alpha+j\omega)\|^2 d\omega & (6.84)\\
&= \sup_{\alpha>0} \frac{1}{2\pi}\int_{-\infty}^{\infty} \bar{\sigma}\left(\mathbf{G}(\alpha+j\omega)\right)^2 \|d(\alpha+j\omega)\|^2 d\omega & (6.85)\\
&\leq \left[\sup_{\alpha>0}\sup_{\omega} \bar{\sigma}\left(\mathbf{G}(\alpha+j\omega)\right)^2\right] \frac{1}{2\pi}\int_{-\infty}^{\infty} \|d(\alpha+j\omega)\|^2 d\omega & (6.86)\\
&= \left[\sup_{\alpha>0}\sup_{\omega} \bar{\sigma}\left(\mathbf{G}(\alpha+j\omega)\right)^2\right] \|d\|_2^2.
\end{aligned}
$$

We use Schwartz's Inequality to get from the first line to the second. The symbol $\bar{\sigma}$ denotes the largest singular value of the matrix transfer function, $\mathbf{G}(\cdot)$. Since $\mathbf{G}$ is a function of the complex number, s, so is $\bar{\sigma}$. Now, since $\|d\|_2^2 < \infty$ by definition, $\|y\|_2^2 < \infty$ if and only if

$$\sup_{\alpha>0}\sup_{\omega} \bar{\sigma}\left(\mathbf{G}(\alpha+j\omega)\right) < \infty. \tag{6.87}$$

The above equation describes the largest possible gain that $\mathbf{G}(s)$ can apply to any possible input, which gives the largest value that $\mathbf{G}$ can obtain. Thus, we define the infinity norm of $\mathbf{G}$ to be

$$\|\mathbf{G}\|_\infty := \sup_{\alpha>0}\sup_{\omega} \bar{\sigma}\Big(\mathbf{G}(\alpha+j\omega)\Big). \tag{6.88}$$

We should note that from our development it is clear that $\|\mathbf{G}\|_\infty$ describes the ratio of the two norms of d and y:

$$\|\mathbf{G}\|_\infty = \frac{\|y\|_2}{\|d\|_2}. \tag{6.89}$$

The body of theory that comprises H_∞ describes the application of the ∞ norm to control problems. Examples of these are the model matching problem and the robust stability and performance problems.

6.3.6 The H_∞ Transfer-Matrix Bound

In this section, the H_∞ norm of the transfer matrix from the disturbance attenuation problem is computed, where it is assumed now that the disturbance inputs of measurement and process noise are L_2 functions. To construct the closed-loop transfer matrix between the disturbance and performance output, the dynamic system coupled to the optimal H_∞ compensator is written together as

$$\begin{aligned}\dot{x}(t) &= Ax(t) + Bu^\circ(t) + \Gamma w(t) = Ax(t) - BR^{-1}B^T S x^\circ(t) + \Gamma w(t),\\ \dot{x}^\circ(t) &= F_c x^\circ(t) + G_c z(t) = F_c x^\circ(t) + G_c H x(t) + G_c v(t),\end{aligned} \tag{6.90}$$

where

$$\begin{aligned}F_c &= A - BR^{-1}B^T S + \theta^{-1}\Gamma W \Gamma^T S - MH^T V^{-1} H,\\ G_c &= MH^T V^{-1}.\end{aligned}$$

Define a new state vector which combines $x(t)$ and $x^\circ(t)$ as

$$\rho(t) = \begin{bmatrix} x(t) \\ x^\circ(t) \end{bmatrix} \tag{6.91}$$

with dynamics system

$$\begin{aligned}\dot{\rho}(t) &= F_{CL}\,\rho(t) + \Gamma_{CL} d(t),\\ y(t) &= C_{CL}\,\rho(t),\end{aligned}$$

where

$$F_{CL} = \begin{bmatrix} A & -B\Lambda \\ G_c H & F_c \end{bmatrix}, \quad d(t) = \begin{bmatrix} w(t) \\ v(t) \end{bmatrix}, \tag{6.92}$$

$$\Gamma_{CL} = \begin{bmatrix} \Gamma & 0 \\ 0 & G_c \end{bmatrix}, \quad C_{CL} = [C \quad -DR^{-1}B^T S]. \tag{6.93}$$

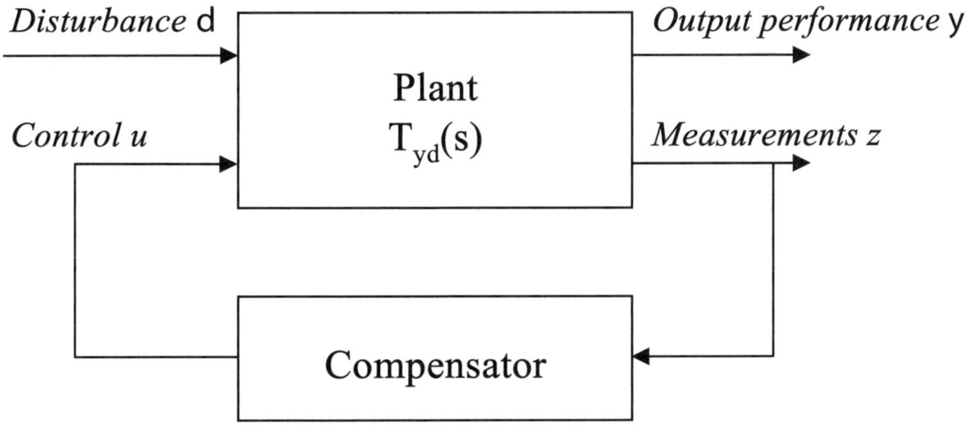

Figure 6.3: Transfer matrix from the disturbance inputs to output performance.

The transfer matrix of the closed-loop system from the disturbances d to the output y is depicted in Figure 6.3.

The transfer matrix T_{yd} is

$$T_{yd}(s) = C_{CL} \left[sI - F_{CL} \right]^{-1} \Gamma_{CL}. \tag{6.94}$$

The following result, proved in [40], shows how the closed-loop transfer matrix is bounded.

Theorem 6.3.2 *The closed-loop system is stable and*

$$\|T_{yd}(s)\|_\infty \leq \theta. \tag{6.95}$$

Example 6.3.1 (Scalar dynamic system) *The characteristics of the ARE and the closed-loop system dynamics are illustrated. Consider the scalar dynamic system*

$$\begin{aligned} \dot{x}(t) &= -1.5x(t) + u(t) + w(t), \\ z(t) &= x(t) + v(t), \end{aligned}$$

where $Q = 4$, $R = 2$, $\theta^{-1} = 1$, $V = 1/14$, $W = 1$, $B = 1$, $\Gamma = 1$, $H = 1$. *The corresponding AREs are*

$$-3S + .5S^2 + 4 = 0 \Rightarrow S = 2,\ 4,$$

$$-3P - 10P^2 + 1 = 0 \Rightarrow P = .2,$$

where we compute $M = 1/3$, $MH^TV^{-1} = 14/3 = G_c$. *A plot of* P *as a function of* θ^{-1} *is shown in Figure* 6.4. *The* $\times$*'s start on the negative reals and continue to decrease as* θ^{-1} *decreases. Then the* $\times$*'s go through* $-\infty$ *to* $+\infty$ *and continue to decreases as* θ^{-1} *decreases till it meets the* $\circ$*'s. At that point it breaks onto the imaginary axis and its solution is no longer valid. At this point the eigenvalues of the Hamiltonian associated with the ARE reach and then split along the imaginary axis if* θ^{-1} *continues to change. Note that there can be two positive solutions. In* [40] *it is shown that only the smallest positive definite solution to the* S *and* P *ARE produces the optimal controller. Here, it is shown that the smallest positive solution to the ARE is associated with the root starting at* $\theta^{-1} = 0$ *or the LQG solution.*

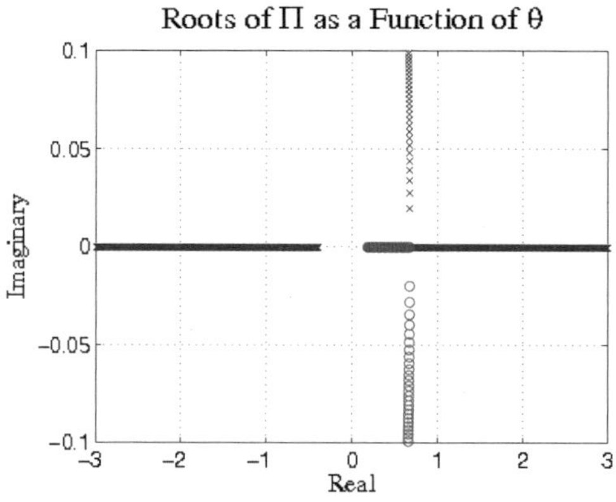

Figure 6.4: Roots of P as a function of θ^{-1}.

The closed-loop matrix (6.92) *for* $S = 2$, $P = .2$,

$$F_{CL} = \begin{bmatrix} -1.5 & -1 \\ 14/3 & -4.2 \end{bmatrix} \Rightarrow \lambda = -2.8 \pm 1.7i, \tag{6.96}$$

where λ is an eigenvalue of F_{CL} and for $S = 4$, $P = .2$,

$$F_{CL} = \begin{bmatrix} -1.5 & -2 \\ 14 & -13.5 \end{bmatrix} \Rightarrow \lambda = -4.7,\ -10.3. \tag{6.97}$$

Note that the complex eigenvalues in (6.96) induced by this approach could not be generated by LQG design for this scalar problem.

Problems

1. Show the results given in Equation (6.48).

2. Find a differential equation for the propagation of $x^\circ(t)$ in Equation (6.46).

3. Consider the system shown in Figure 6.5 Assume $t_f \to \infty$ and all parameters are time invariant. Assume

$$\begin{aligned} \dot{S} &\to 0, \\ \dot{P} &\to 0, \end{aligned}$$

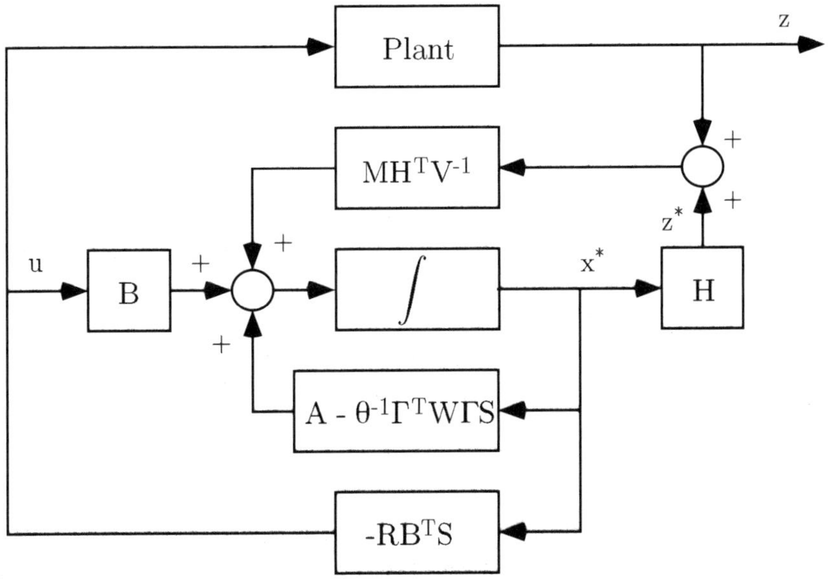

Figure 6.5: System description.

which means using the ARE. The system equations are

$$\begin{aligned}\dot{x}(t) &= ax(t) + u + w(t),\\ z(t) &= x(t) + v(t),\\ y &= \begin{bmatrix} 1 \\ 0 \end{bmatrix} x(t) + \begin{bmatrix} 0 \\ 1 \end{bmatrix} u,\end{aligned} \tag{6.98}$$

where

$$\begin{aligned} Q = 1, \quad R = 1,\\ W = 1, \quad V = 1. \end{aligned} \tag{6.99}$$

(a) Plot S as a function of θ^{-1} for $a = 1$ and $a = -1$.

(b) Plot P as a function of θ^{-1} for $a = 1$ and $a = -1$.

(c) For some choice of θ^{-1} show that all necessary conditions are satisfied:

i. $P > 0$,

ii. $S \geq 0$,

iii. $I - \theta^{-1} PS > 0$.

(d) Write down the compensator.

Appendix A
Background

A.1 Topics from Calculus

In this section we present some results from calculus that will be of use in the main text. Note that a basic familiarity with such topics as continuity, differentiation, and integration is assumed.

A.1.1 Implicit Function Theorems

It should be noted that this section is adapted from Friedman's book *Advanced Calculus* [21].

Let $F(x, y)$ be a function defined in an open set G of $\mathbb{R}^2$. We are concerned with the set of points (x, y) in G satisfying $F(x, y) = 0$. Can we write the points of this set in the form $y = g(x)$? If so, then we say that the equation $F(x, y) = 0$ defines the function $y = g(x)$ *implicitly*, and we call the equation

$$F(x, y) = 0 \tag{A.1}$$

the *implicit equation* for $y = g(x)$.

If $F(x, y) = x^2 + y^2$ in $\mathbb{R}^2$, then the only solution of Equation (A.1) is $(0, 0)$. Thus there is no function $y = g(x)$ that is defined on some interval of positive length.

If $F(x, y) = x^2 + y^2 - 1$ in $\mathbb{R}^2$, then there are two functions $y = g(x)$ for which $(x, g(x))$ satisfies Equation (A.1) for $-1 \leq x \leq 1$, namely,

$$y = g(x) = \pm\sqrt{1 - x^2}. \tag{A.2}$$

Note, however, that if we restrict (x, y) to belong to a small neighborhood of a solution (x_0, y_0) of Equation (A.1) with $|x_0| < 1$, then there is a unique solution $y = g(x)$ satisfying $g(x_0) = y_0$.

We shall now prove a general theorem asserting that the solutions (x, y) of Equation (A.1) in a small neighborhood of a point (x_0, y_0) have the form $y = g(x)$. This theorem is called the *Implicit Function Theorem* for a function of two variables.

Theorem A.1.1 *Let $F(x, y)$ be a function defined in an open set G of $\mathbb{R}^2$ having continuous first derivatives, and let (x_0, y_0) be a point of G for which*

$$F(x_0, y_0) = 0, \tag{A.3}$$

$$F_y(x_0, y_0) \neq 0. \tag{A.4}$$

Then there exists a rectangle R in G defined by

$$|x - x_0| < \alpha, \qquad |y - y_0| < \beta \tag{A.5}$$

such that the points (x, y) in R that satisfy Equation (A.1) *have the form* $(x, g(x))$, *where $g(x)$ is a function having a continuous derivative in $|x - x_0| < \alpha$. Furthermore,*

$$g'(x) = -\frac{F_x(x, g(x))}{F_y(x, g(x))} \quad \text{if } |x - x_0| < \alpha. \tag{A.6}$$

The condition (A.4) cannot be omitted, as shown by the example of $F(x, y) = x^2 + y^2$, $(x_0, y_0) = (0, 0)$.

Proof: We may suppose that $F_y(x_0, y_0) > 0$. By continuity, $F_y(x, y) > 0$ if $|x - x_0| \leq d$, $|y - y_0| \leq d$, where d is a small positive number. Consider the function $\phi(y) = F(x_0, y)$. It is strictly monotone increasing since $\phi'(y) = F_y(x_0, y) > 0$. Since $\phi(y_0) = F(x_0, y_0) = 0$, it follows that

$$F(x_0, y_0 - d) = \phi(y_0 - d) < 0 < \phi(y_0 + d) = F(x_0, y_0 + d). \tag{A.7}$$

Using the continuity of $F(x, y_0 - d)$ and of $F(x, y_0 + d)$ we deduce that

$$F(x, y_0 - d) < 0, \quad F(x, y_0 + d) > 0 \tag{A.8}$$

if $|x - x_0|$ is sufficiently small, say, if $|x - x_0| \leq d_1$. See Figure A.1.

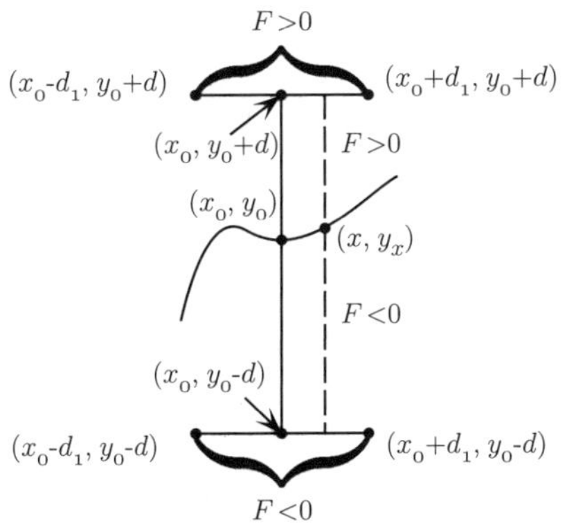

Figure A.1: Definition of $F_y(x_0, y_0) > 0$.

Consider, in the interval $|y - y_0| \leq d$, the continuous function $\psi(y) = F(x, y)$ for x fixed, $|x - x_o| \leq d_2$, $d_2 = \min(d, d_1)$. It is strictly monotone increasing since $\psi_y = F_y(x, y) > 0$. Also, by Equation (A.8), $\psi(y_0 - d) < 0$, $\psi(y_0 + d) > 0$. From the continuity of $\psi(\cdot)$, there is a point y_x in the interval $(y_o - d, y_0 + d)$ satisfying $\psi(y_x) = 0$. Since ψ is strictly monotone, y_x is unique. Writing $y_x = g(x)$, we have proved that the solutions of Equation (A.1) for $|x - x_0| < d_2$, $|y - y_0| < d$ have the form $(x, g(x))$.

We shall next prove that $g(x)$ is continuous. For any $\epsilon > 0$, $\epsilon \leq d$, $F(x_0, y_0+\epsilon) > 0 > F(x_0, y_0 - \epsilon)$. Repeating the argument given above with $y_0 \pm d$ replaced by $y_0 \pm \epsilon$, we see that there exists a number $d_2(\epsilon)$ such that if $|x - x_0| < d_2(\epsilon)$, then there is a unique $y = \bar{y}_x$ in $(y_0 - \epsilon, y_0 + \epsilon)$ satisfying $F(x, \bar{y}_x) = 0$. By uniqueness of the solution y of $F(x, y) = 0$ in $(y_0 - d, y_0 + d)$, it follows that $\bar{y}_x = g(x)$. Hence

$$|g(x) - y_0| < \epsilon \qquad \text{if } |x - x_0| \leq d_2(\epsilon). \tag{A.9}$$

This proves the continuity of $g(x)$ at x_0.

Now let x_1 be any point in $|x - x_0| < d_2$. Then $F(x_1, y_1) = 0$, where $y_1 = g(x_1)$, and $F_y(x1, y1) > 0$. We therefore can apply the proof of the continuity of $g(x)$ at x_0 and deduce the continuity of $g(x)$ at x_1.

We proceed to prove that $g(x)$ is differentiable. We begin with the relation

$$F(x_0 + h, g(x_0 + h)) - F(x_0, g(x_0)) = 0, \tag{A.10}$$

where $|h| < d_2$. Writing $g(x_0 + h) = g(x_0) + \Delta g$ and using the differentiability of F, we get

$$hF_x(x_0, y_0) + \Delta g \cdot F_y(x_0, y_0) + \eta\sqrt{h^2 + (\Delta g)^2} = 0, \tag{A.11}$$

where $\eta = \eta(h, \Delta g) \to 0$ if $(h, \Delta g) \to 0$. Since $g(x)$ is continuous, $\Delta g \to 0$ if $h \to 0$. Hence $\eta \to 0$ if $h \to 0$.

Writing $F_x(x_0, y_0) = F_x$, $F_y(x_0, y_0) = F_y$, and dividing both sides of Equation (A.11) by hF_y, we find that

$$\frac{\Delta g}{h} = -\frac{F_x}{F_y} + \frac{\eta}{hF_y}\sqrt{h^2 + (\Delta g)^2}. \tag{A.12}$$

If $|h|$ is sufficiently small, $|\eta/F_y| < 1/2$, hence

$$\frac{|\Delta g|}{|h|} \leq \frac{|F_x|}{|F_y|} + \frac{1}{2}\frac{|\Delta g|}{|h|}.$$

It follows that

$$\frac{|\Delta g|}{|h|} \leq C, \qquad C \text{ constant}.$$

Using this in Equation (A.12) and taking $h \to 0$, we conclude that

$$\lim_{h\to 0} \frac{\Delta g}{h} \tag{A.13}$$

exists and is equal to $-F_x/F_y$.

We have thus proved that $g'(x)$ exists at x_0 and that Equation (A.6) holds at x_0. The same argument can be applied at any point x in $|x - x_0| < d_2$. Thus $g'(x)$ exists and it satisfies Equation (A.6). Since the right-hand side of Equation (A.6) is continuous, the same is true of $g'(x)$. This completes the proof of the theorem. ∎

If instead of Equation (A.4) we assume that $F_x(x_0, y_0) \neq 0$, then we can prove an analogue of Theorem A.1.1 with the roles of x and y interchanged. The proof of Theorem A.1.1 extends to the case where $x = (x_1, x_2, \ldots, x_n)$. The result is the following *implicit function theorem* for a function of several variables.

Theorem A.1.2 *Let $F(x, y) = F(x_1, x_2, \ldots, x_n, y)$ be a function defined in an open set of G of $\mathbb{R}^{n+1}$ and let $(x^0, y^0) = (x_1^0, \ldots, x_n^0, y^0)$ be a point of G. Assume that*

$$F(x^0, y^0) = 0, \qquad F_y(x^0, y^0) \neq 0. \tag{A.14}$$

Then there exists an $(n+1)$-dimensional rectangle R defined by

$$|x_i - x_i^0| < \alpha (1 \leq i \leq n), \qquad |y - y^0| < \beta \tag{A.15}$$

such that for any x in Equation (A.15) *there exists a unique solution $y = y_x$ of $F(x, y) = 0$ in the interval $|y - y^0| < \beta$. Writing $y_x = g(x)$, the function $g(x)$ is continuously differentiable, and*

$$\frac{\partial}{\partial x_i} g(x) = -\frac{(\partial F/\partial x_i)(x, g(x))}{(\partial F/\partial y)(x, g(x))} \qquad (1 \le i \le n). \tag{A.16}$$

Consider next a more complicated situation in which we want to solve two equations simultaneously:

$$F(x, y, z, u, v) = 0, \tag{A.17}$$

$$G(x, y, z, u, v) = 0. \tag{A.18}$$

We introduce the determinant

$$J = \left| \frac{\partial(F, G)}{\partial(u, v)} \right| = \left| \begin{array}{cc} F_u & F_v \\ G_u & G_v \end{array} \right| \tag{A.19}$$

called the *Jacobian* of F, G with respect to u, v.[13]

Theorem A.1.3 *Let F and G have continuous first derivatives in an open set D of $\mathbb{R}^5$ containing a point $P_0 = (x_0, y_0, z_0, u_0, v_0)$. Assume that*

$$F(x_0, y_0, z_0, u_0, v_0) = 0, \qquad G(x_0, y_0, z_0, u_0, v_0) = 0, \tag{A.20}$$

and

$$\left. \frac{\partial(FG)}{\partial(u, v)} \right|_{(x_0, y_0, z_0, u_0, v_0)} \neq 0, \tag{A.21}$$

which means the Jacobian is nonzero. Then there exists a cube

$$R : |x - x_0| < \alpha, \qquad |y - y_0| < \alpha, \qquad |z - z_0| < \alpha, \tag{A.22}$$

and a rectangle

$$S : |u - u_0| < \beta_1, \qquad |v - v_0| < \beta_2 \tag{A.23}$$

[13]The Jacobian *determinant* should not be confused with the Jacobian matrix from Definition A.2.4.

such that for any (x, y, z) *in* R *there is a unique pair* (u, v) *in* S *for which Equation* (A.17) *and* (A.18) *hold. Writing*

$$u = f(x, y, z), \qquad v = g(x, y, z), \tag{A.24}$$

the functions f *and* g *have continuous first derivatives in* R, *and*

$$f_x = -\frac{1}{J}\frac{\partial(F,G)}{\partial(x,v)} = -\frac{1}{J}\begin{vmatrix} F_x & F_v \\ G_x & G_v \end{vmatrix}, \tag{A.25}$$

$$g_x = -\frac{1}{J}\frac{\partial(F,G)}{\partial(u,x)} = -\frac{1}{J}\begin{vmatrix} F_u & F_x \\ G_u & G_x \end{vmatrix}. \tag{A.26}$$

Similar formulas hold for f_y, f_z, g_y, g_z.

Proof: Since $J \neq 0$ at P_0, either $F_v \neq 0$ or $G_v \neq 0$ at P_0. Suppose $F_v \neq 0$ at P_0. By Theorem A.1.2, if (x, y, z) lies in a small rectangle T with center (x_0, y_0, z_0, u_0), then there exists a unique solution $p = \phi(x, y, z, u)$ of Equation (A.17) in some small interval $|v - v_0| < \beta_2$. Let

$$H(x, y, z, u) = G(x, y, z, u, \phi(x, y, z, u)). \tag{A.27}$$

Then (u, v) is a solution of Equations (A.17) and (A.18) (when $(x, y, z, u) \in T$, $|v - v_0| < \beta_2$) if and only if $v = \phi(x, y, z, u)$ and

$$H(x, y, z, u) = 0. \tag{A.28}$$

ϕ has continuous first derivatives and $\phi_u = -F_u/F_v$. Hence

$$H_u = G_u + G_v\phi_u = G_u - \frac{G_vF_u}{F_v} = \frac{G_uF_v - G_vF_u}{F_v} = -\frac{J}{F_v} \neq 0 \tag{A.29}$$

at P_0. We therefore can apply Theorem A.1.2. We conclude that for any (x, y, z) in a small cube R with center (x_0, y_0, z_0) there is a unique solution of u of Equation (A.28) in some interval $|u - u_0| < \beta_1$; the points (x, y, z, u) belong to T. Furthermore, this solution u has the form

$$u = g(x, y, z), \tag{A.30}$$

where g has continuous first derivatives. It follows that

$$v = \phi(x, y, z, g(x, y, z)) \tag{A.31}$$

also has continuous first derivatives.

It remains to prove Equations (A.25) and (A.26). To do this we differentiate the equations

$$\begin{aligned} F(x, y, z, f(x, y, z), g(x, y, z)) &= 0, \\ G(x, y, z, f(x, y, z), g(x, y, z)) &= 0 \end{aligned}$$

with respect to x and get

$$F_x + F_u f_x + F_v g_x = 0, \qquad G_x + G_u f_x + G_v g_x = 0. \tag{A.32}$$

Solving for f_x, g_x, we get Equations (A.25) and (A.26). ∎

We conclude this section with a statement of the most general implicit function theorem for a system of functions. Let

$$F_i(x, u) = F_i(x_1, \ldots, x_n, u_1, \ldots, u_r) \qquad (1 \le i \le r) \tag{A.33}$$

be functions having continuous first derivatives in an open set containing a point (x^0, u^0). The matrix

$$\begin{bmatrix} \frac{\partial F_1}{\partial u_1} & \frac{\partial F_1}{\partial u_2} & \cdots & \frac{\partial F_1}{\partial u_r} \\ \frac{\partial F_2}{\partial u_1} & \frac{\partial F_2}{\partial u_2} & \cdots & \frac{\partial F_2}{\partial u_r} \\ \vdots & \vdots & \cdots & \vdots \\ \frac{\partial F_r}{\partial u_1} & \frac{\partial F_r}{\partial u_2} & \cdots & \frac{\partial F_r}{\partial u_r} \end{bmatrix} \tag{A.34}$$

or briefly, $(\partial F_i/\partial u_j)$ is called the *Jacobian matrix* of $(F_1, \ldots, F_r)$ with respect to $(u_1, \ldots, u_r)$. The determinant of this matrix is called the *Jacobian* of $(F_1, \ldots, F_r)$

with respect to $(u_1, \ldots, u_r)$ and is denoted by

$$J \triangleq \frac{\partial(F_1, \ldots, F_r)}{\partial(u_1, \ldots, u_r)}. \tag{A.35}$$

Theorem A.1.4 *Let $F_1, \ldots, F_r$ have continuous first derivatives in a neighborhood of a point (x^0, u^0). Assume that*

$$F_i(x^0, u^0) = 0 \quad (1 \le i \le r), \qquad \frac{\partial(F_1, \ldots, F_r)}{\partial(u_1, \ldots, u_r)} \neq 0 \quad at \ (x^0, u^0). \tag{A.36}$$

Then there is a δ-neighborhood R of x^0 and a γ-neighborhood S of u^0 such that for any x in R there is a unique solution u of

$$F_i(x, u) = 0 \qquad (1 \le i \le r) \tag{A.37}$$

in S. The vector valued function $u(x) = (u_1(x), \ldots, u_r(x))$ thus defined has continuous first derivatives in R.

In order to compute $\partial u_i / \partial x_j$ $(1 \le i \le r)$ for a fixed j, we differentiate the equations

$$F_1(x, u(x)) = 0, \ldots, F_r(x, u(x)) = 0 \tag{A.38}$$

with respect to x_j. We obtain the system of linear equations for $\partial u_i / \partial x_j$:

$$\frac{\partial F_k}{\partial x_j} + \sum_{i=1}^{r} \frac{\partial F_k}{\partial u_i} \frac{\partial u_i}{\partial x_j} = 0 \qquad (1 \le k \le r). \tag{A.39}$$

The system of linear equations (A.39) in the unknowns $\partial u_i / \partial x_j$ can be uniquely solved, since the determinant of the coefficients matrix, which is precisely the Jacobian

$$\frac{\partial(F_1, \ldots, F_r)}{\partial(u_1, \ldots, u_r)},$$

is different from 0.

We briefly give the proof of Theorem A.1.4. It is based upon induction on r. Without loss of generality we may assume that

$$\frac{\partial(F_1,\ldots,F_{r-1})}{\partial(u_1,\ldots,u_{r-1})} \neq 0. \tag{A.40}$$

Therefore, by the inductive assumption, the solution of

$$F_i(x,u) = 0 \qquad (1 \leq i \leq r-1) \tag{A.41}$$

in a neighborhood of (x^0, u^0) is given by $u_i = \phi_i(x, u_r)$ with $(1 \leq i \leq r-1)$. Let

$$G(x,u_r) = F_r(x, \phi_1(x,u_r), \ldots, \phi_{r-1}(x,u_r), u_r). \tag{A.42}$$

If we show that

$$\frac{\partial G}{\partial u_r} \neq 0 \qquad \text{at } (x^0, u^0), \tag{A.43}$$

then we can use Theorem A.1.2 to solve the equation $G(x,u_r) = 0$. To prove Equation (A.43), differentiate the equations

$$F_i(x, \phi_1(x,u_r), \ldots, \phi_{r-1}(x,u_r), u_r) = 0 \qquad (1 \leq i \leq r-1) \tag{A.44}$$

with respect to u_r to obtain

$$\sum_{j=1}^{r-1} \frac{\partial F_1}{\partial u_j}\frac{\partial \phi_j}{\partial u_r} + \frac{\partial F_i}{\partial u_r} = 0 \qquad (1 \leq i \leq r-1). \tag{A.45}$$

Differentiate also Equations (A.42) with respect to u_r to obtain

$$\sum_{j=1}^{r-1} \frac{\partial F_r}{\partial u_j}\frac{\partial \phi_j}{\partial u_r} - \frac{\partial G}{\partial u_r} + \frac{\partial F_r}{\partial u_r} = 0. \tag{A.46}$$

Solving the linear system of Equation (A.45) and (A.46) (in the unknowns $\partial\phi_j/\partial u_r$, $\partial G/\partial u_r$) for $\partial G/\partial u_r$, we obtain

$$\frac{\partial G}{\partial u_r} = \frac{\partial(F_1,\ldots,F_r)/\partial(u_1,\ldots,u_r)}{\partial(F_1,\ldots,F_{r-1})/\partial(u_1,\ldots,u_{r-1})}. \tag{A.47}$$

This gives Equation (A.43).

A.1.2 Taylor Expansions

We first introduce the definition of order.

Definition A.1.1 *Consider a function $R(x)$. We say that $R(\cdot)$ is* of higher order *than x^n and write*

$$R(x) \sim \mathcal{O}(x^n)$$

if

$$\lim_{x\to 0} \frac{R(x)}{x^n} = \lim_{x\to 0} \frac{\mathcal{O}(x^n)}{x^n} = 0.$$

Taylor's Expansion for Functions of a Single Variable

We begin by reviewing the definition and the simplest properties of the Taylor expansion for functions of one variable. If $f(x)$ has an Nth derivative at x_0 its *Taylor expansion of degree N about x_0* is the polynomial

$$f(x_0) + \frac{1}{1!}f'(x_0)(x-x_0) + \frac{1}{2!}f''(x_0)(x-x_0)^2 + \cdots + \frac{1}{N!}f^{(N)}(x_0)(x-x_0)^N. \quad \text{(A.48)}$$

The relation between f and its Taylor expansion can be expressed conveniently by the following *integral remainder formula.*

Theorem A.1.5 *If f has a continuous Nth derivative in a neighborhood of x_0, then in that neighborhood*

$$f(x) = f(x_0) + \frac{1}{1!}f'(x_0)(x-x_0) + \cdots + \frac{1}{N!}f^{(N)}(x_0)(x-x_0)^N + R_N, \quad \text{(A.49)}$$

where

$$R_N = \frac{1}{(N-1)!}\int_{x_0}^{x} (x-t)^{N-1}\left[f^{(N)}(t) - f^{(N)}(x_0)\right] dt, \quad \text{(A.50)}$$

where R_N is of order $\mathcal{O}((x-x_0)^N)$ and

$$\lim_{(x-x_0)\to 0} \frac{\mathcal{O}((x-x_0)^N)}{|(x-x_0)^N|} \to 0. \quad \text{(A.51)}$$

Proof: The remainder can be written as the difference

$$R_N = \frac{1}{(N-1)!}\int_{x_0}^{x}(x-t)^{N-1}f^{(N)}(t)dt - \frac{f^{(N)}(x_0)}{(N-1)!}\int_{x_0}^{x}(x-t)^{N-1}dt. \quad (A.52)$$

The second of these integrals is directly computed to be

$$\frac{f^{(N)}(x_0)}{(N-1)!}\int_{x_0}^{x}(x-t)^{N-1}dt = \frac{1}{N!}f^{(N)}(x_0)(x-x_0)^N, \quad (A.53)$$

which is just the last term of the Taylor expansion. The first integral can be integrated by parts, which together with (A.53) leads to

$$\frac{1}{(N-2)!}\int_{x_0}^{x}(x-t)^{N-2}[f^{(N-1)}(t) - f^{(N-1)}(x_0)]dt = R_{N-1}. \quad (A.54)$$

We therefore obtain

$$R_N = -\frac{1}{N!}f^{(N)}(x_0)(x-x_0)^N + R_{N-1}. \quad (A.55)$$

If we substitute the preceding equation into Equation (A.49), we get Equation (A.49) back again with N replaced by $N-1$. The induction is completed by noticing that, for $N=1$, Equation (A.49) is just

$$f(x) = f(x_0) + f'(x_0)(x-x_0) + \int_{x_0}^{x}[f'(t) - f'(x_0)]dt \quad (A.56)$$

and that this is a valid equation. Finally, the remainder R_N in (A.50) is shown to be $\mathcal{O}(x-x_0)^N$. The following inequality can be constructed:

$$\begin{aligned} |R_N| &= \left|\frac{1}{(N-1)!}\int_{x_0}^{x}(x-t)^{N-1}\left[f^{(N)}(t) - f^{(N)}(x_0)\right]dt\right| \\ &\leq \max_{x' \in (x-x_0)}\left[|f^{(N)}(x') - f^{(N)}(x_0)|\right]\frac{1}{(N-1)!}\left|\int_{x_0}^{x}(x-t)^{N-1}dt\right| \\ &= \max_{x' \in (x-x_0)}\left[|f^{(N)}(x') - f^{(N)}(x_0)|\right]\frac{1}{N!}|(x-x_0)|^N. \end{aligned} \quad (A.57)$$

Since it is assumed that $f^{(N)}$ is continuous, then using (A.57)

$$\lim_{(x-x_0)\to 0} \frac{R_N}{|(x-x_0)^N|} \to 0, \tag{A.58}$$

which by Definition A.1.1 implies that R_N is $\mathcal{O}((x-x_0)^N)$. ∎

Remark A.1.1 *A vector representation for Taylor's Expansion is given in Section* A.2.10, *after the linear algebra review of Appendix* A.2.

A.2 Linear Algebra Review

It is assumed that the reader is familiar with the concepts of vectors, matrices, and inner products. We offer here a brief review of some of the definitions and concepts that are of particular interest to the material in the main text.

A.2.1 Subspaces and Dimension

A subspace $\mathcal{S}$ of $\mathbb{R}^n$ is a set of vectors in $\mathbb{R}^n$ such that for any $s_1, s_2 \in \mathcal{S}$ and any two scalars $\alpha, \beta \in \mathbb{R}$, then

$$\alpha s_1 + \beta s_2 \in \mathcal{S}.$$

The span of $\mathcal{S} \subset \mathbb{R}^n$ is defined as the collection of all finite linear combinations of the elements of $\mathcal{S}$. Suppose we have a set of vectors $\{e_1, e_2, \ldots, e_m\}$ in $\mathbb{R}^n$. Then the set of vectors

$$\mathcal{S} = \{s \mid s = \alpha_1 e_1 + \alpha_2 e_2 + \cdots + \alpha_m e_m, \alpha_i \in \mathbb{R}\}$$

(read, "the set of all s such that $s = \alpha_1 e_1 + \cdots$") is a subspace and is spanned by $\mathcal{S}$. A set of vectors $\{e_1, e_2, \ldots, e_m\}$ in $\mathbb{R}^n$ is said to be *linearly independent* if

$$\alpha_1 e_1 + \alpha_2 e_2 + \cdots + \alpha_m e_m = 0 \Leftrightarrow \alpha_1 = \alpha_2 = \cdots = \alpha_m = 0. \tag{A.59}$$

The set of vectors $\{e_1, e_2, \ldots, e_m\}$ is a *basis* for the subspace if they span $\mathcal{S}$ and are independent.

The *dimension* of a subspace is the smallest number of vectors required to form a basis. It can be shown that any set of m linearly independent vectors in the subspace is a basis for the space. That is, if $\{s_1, s_2, \ldots, s_m\}$ are linearly independent vectors in the m-dimensional subspace $\mathcal{S}$, then any $s \in \mathcal{S}$ can be written as

$$s = \alpha_1 s_1 + \alpha_2 s_2 + \cdots + \alpha_m s_m,$$

where $\alpha_i, i = 1, 2, \ldots m$, are real scalars.

Example A.2.1 *Consider the vectors*

$$e_1 = \begin{Bmatrix} 1 \\ 1 \\ 2 \end{Bmatrix}, \quad e_2 = \begin{Bmatrix} 1 \\ 0 \\ 1 \end{Bmatrix}, \quad e_3 = \begin{Bmatrix} 0 \\ 1 \\ 1 \end{Bmatrix}.$$

These form a basis for a subspace $\mathcal{S} \subset \mathbb{R}^3$. *However, the vectors are not linearly independent, as*

$$e_1 - e_2 - e_3 = 0,$$

so that the requirement of (A.59) *above is not satisfied. It can be seen that at most two of the three vectors are linearly independent, so that* $\mathcal{S}$ *has dimension* 2. *Any two linearly independent vectors in* $\mathcal{S}$ *can serve as a basis. In particular, both*

$$\{e_1, e_2\} \qquad \text{and} \qquad \{e_2, e_3\}$$

are bases for the space.

A.2.2 Matrices and Rank

Given a real matrix $A \in \mathbb{R}^{m \times n}$, the *range space* of the matrix is the set of all vectors $y \in \mathbb{R}^m$ that can be written as Ax, where $x \in \mathbb{R}^n$. That is,

$$\text{Range}(A) = \{y \mid y = Ax \text{ for some } x \in \mathbb{R}^n\}.$$

The *null space* of A is the set of vectors that when multiplied by A produce zero:

$$\text{Null}(A) = \{w \mid 0 = Aw\}.$$

The *rank* of the matrix A is the dimension of the range space. It is obvious from the definition that the largest possible rank of an m by n matrix is the smaller of m and n.

For a square matrix $A \in \mathbb{R}^{n \times n}$, the dimensions of the range and null spaces sum to n.

The *inverse* of the square matrix A is the matrix A^{-1} such that

$$AA^{-1} = I.$$

The inverse exists if and only if the matrix is of full rank, that is, the rank of the $n \times n$ matrix is n. A matrix for which the inverse does not exist is called *singular*.

A.2.3 Minors and Determinants

Here we make some statements about the determinants of square matrices. For a more complete treatment, see, for instance, Cullen [16].

For a matrix with a single element, the determinant is declared to be the value of the element. That is, let a_{11} be the only element of the 1×1 matrix A. Then the determinant $|A|$ of A is

$$|A| = a_{11}.$$

Laplace Expansion

Let M_{ij} be the matrix created by deleting row i and column j from A. Then the determinant can be computed from the Laplace expansion as

$$|A| = \sum_{j=1}^{n} a_{ij}(-1)^{i+j}|M_{ij}|$$

for any row i. The expansion also holds over columns, so that the order of the subscripts in the expansion can be reversed and the summation taken over any column in the matrix.

The value $C_{ij} \triangleq (-1)^{i+j}|M_{ij}|$ is generally called the *cofactor* of the element a_{ij}. The determinant $|M_{ij}|$ is known as the *minor* of element a_{ij}.[14]

Trivially, the Laplace expansion for a matrix of dimension 2 is

$$\begin{vmatrix} a_{11} & a_{12} \\ a_{21} & a_{22} \end{vmatrix} = a_{11}a_{22} - a_{12}a_{21},$$

where we have used the definition of the determinant of a 1×1 matrix to evaluate the minors.

The cofactors for matrix of dimension 3 involve matrices of dimension 2, so this result can be used, along with the Laplace expansion, to compute the determinant for a 3×3 matrix, etc.

The determinant has several useful properties. Among these are the following:

1. $|AB| = |A||B|$.
2. If A is $n \times n$, then $|A| = 0$ if and only if the rank of A is less than n. That is, $|A| = 0$ is identical to saying that A is singular.
3. $|A^{-1}| = 1/|A|$.

A.2.4 Eigenvalues and Eigenvectors

An *eigenvalue* of a square matrix A is a scalar λ (in general complex) such that the determinant

$$|A - \lambda I| = 0, \tag{A.60}$$

[14]Some authors refer to the matrix M_{ij} itself as the minor, rather than its determinant.

where I is the appropriately dimensioned identity matrix. Equation (A.60) is called the *characteristic equation* of the matrix; the values of λ for which it is satisfied are sometimes known as the *characteristic roots* or *characteristic values* of A, as well as eigenvalues.

An *eigenvector* is a vector $v \neq 0$ such that

$$Av = \lambda v$$

or, equivalently,

$$[A - \lambda I]v = 0.$$

Clearly, v can exist only if $|A - \lambda I| = 0$, so that λ is an eigenvalue of the matrix. Note that even for real A, the eigenvectors are in general complex.

If v is an eigenvector of A, then αv is also an eigenvector (corresponding to the same eigenvalue) for all $\alpha \in \mathbb{C}$.

A.2.5 Quadratic Forms and Definite Matrices

This section introduces certain definitions and concepts which are of fundamental importance in the study of networks and systems. Some of what follows is adapted from the book *Systems, Networks, and Computation* by Athans et al. [3].

Suppose that x is a column n-vector with components x_i and that A is a real $n \times n$ symmetric matrix, that is, $A = A^T$, with elements a_{ij}. Let us consider the scalar-valued function $f(x)$ defined by the scalar product

$$f(x) = \langle x, Ax \rangle = \sum_{i=1}^{n} \sum_{j=1}^{n} a_{ij} x_i x_j. \tag{A.61}$$

This is called a *quadratic form* because it involves multiplication by pairs of the elements x_i of x. For example, if

$$x = \begin{bmatrix} x_1 \\ x_2 \end{bmatrix}, \quad A = \begin{bmatrix} 1 & 2 \\ 2 & 3 \end{bmatrix}, \tag{A.62}$$

then

$$f(x) = \langle x, Ax \rangle = x_1^2 + 4x_1x_2 + 3x_2^2, \tag{A.63}$$

which involves terms in the square of the components of x and their cross products.

We now offer certain definitions.

Definition A.2.1 *If for all x,*

(a)
$$f(x) = \langle x, Ax \rangle \geq 0,$$

$f(x)$ is called a nonnegative definite form and A is called a nonnegative definite matrix;

(b)
$$f(x) = \langle x, Ax \rangle > 0,$$

$f(x)$ is called a positive definite form and A is called a positive definite matrix;

(c)
$$f(x) = \langle x, Ax \rangle \leq 0,$$

$f(x)$ is called a nonpositive definite form and A is called a nonpositive definite matrix;

(d)
$$f(x) = \langle x, Ax \rangle < 0,$$

$f(x)$ is called a negative definite form and A is called a negative definite matrix.

We now give a procedure for testing whether a given matrix is positive definite. The basic technique is summarized in the following theorem.

Theorem A.2.1 *Suppose that A is the real symmetric $n \times n$ matrix*

$$A = \begin{bmatrix} a_{11} & a_{12} & \cdots & a_{1n} \\ a_{12} & a_{22} & \cdots & a_{2n} \\ \vdots & \vdots & \vdots & \vdots \\ a_{1n} & a_{2n} & \cdots & a_{nn} \end{bmatrix}. \tag{A.64}$$

Let A_k be the $k \times k$ matrix, defined in terms of A, for $k = 1, 2, \ldots, n$, by

$$A_k = \begin{bmatrix} a_{11} & a_{12} & \cdots & a_{1k} \\ a_{12} & a_{22} & \cdots & a_{2k} \\ \vdots & \vdots & \vdots & \vdots \\ a_{1k} & a_{2k} & \cdots & a_{kk} \end{bmatrix}. \tag{A.65}$$

Then A is positive definite if and only if

$$\det\ A_k > 0 \tag{A.66}$$

for each $k = 1, 2, \ldots, n$.

There is a host of additional properties of definite and semidefinite symmetric matrices, which we give below as theorems. Some of the proofs are easy, but others are very difficult. Suppose that A is a real symmetric matrix of dimension n. The characteristic value problem is to determine the scalar λ and the nonzero vectors $v \in \mathbb{R}^n$ which simultaneously satisfy the equation

$$Av = \lambda v \quad \text{or} \quad (A - \lambda I)v = 0. \tag{A.67}$$

This system of n linear equations in the unknown vector v has a nontrivial solution if and only if $\det(A - \lambda I) = 0$, the characteristic equation.

Theorem A.2.2 *The characteristic roots or eigenvalues of a symmetric matrix are all real.*

Proof: Let A be a symmetric matrix and let λ be any root of the characteristic equation. Then

$$Av = \lambda v, \tag{A.68}$$

where v, the eigenvector, may be a complex vector. The conjugate transpose of v is denoted v^*. Then

$$v^* Av = \lambda v^* v. \tag{A.69}$$

Since v^*Av is a scalar, and A is real, i.e., $A^* = A$,

$$(v^*Av)^* = (v^*Av); \tag{A.70}$$

that is, v^*Av satisfies its own conjugate and hence must be real. Since v^*Av and v^*v are real, λ must be real. ∎

Theorem A.2.3 *For a symmetric matrix, all the n vectors v associated with the n eigenvalues λ are real.*

Proof: Since $(A - \lambda I)$ is real, the solution v of $(A - \lambda I)v = 0$ must be real. ∎

Theorem A.2.4 *If v_1 and v_2 are eigenvectors associated with the distinct eigenvalues λ_1 and λ_2 of a symmetric matrix A, then v_1 and v_2 are orthogonal.*

Proof: We know that $Av_1 = \lambda_1 v_1$ and $Av_2 = \lambda_2 v_2$. This implies that ${v_2}^T Av_1 = \lambda_1 {v_2}^T v_1$ and ${v_1}^T Av_2 = \lambda_2 {v_1}^T v_2$. Taking the transpose of the first equation gives

$$v_1^T Av_2 = \lambda_1 v_1^T v_2. \tag{A.71}$$

Subtract the second equation to obtain

$$(\lambda_1 - \lambda_2)(v_1^T v_2) = 0. \tag{A.72}$$

Since $\lambda_1 - \lambda_2 \neq 0$, then ${v_1}^T v_2 = 0$. ∎

Suppose all the eigenvalues are distinct. Then

$$V = [v_1, \ldots, v_n]^T \tag{A.73}$$

is an orthogonal matrix. This means that since

$$V^T V = I, \tag{A.74}$$

then

$$V^T = V^{-1}. \tag{A.75}$$

Even if the eigenvalues are repeated, the eigenmatrix V is still orthogonal [27]. Therefore,

$$AV = VD, \tag{A.76}$$

where D is a diagonal matrix of the eigenvalues $\lambda_1, \ldots, \lambda_n$. Therefore,

$$D = V^T AV, \tag{A.77}$$

where V is the orthogonal matrix forming the similarity transformation.

Theorem A.2.5 *A is a positive definite matrix if and only if all its eigenvalues are positive. A is a negative definite matrix if and only if all its eigenvalues are negative. In either case, the eigenvectors eigenvector of A are real and mutually orthogonal.*

Theorem A.2.6 *If A is a positive semidefinite or negative semidefinite matrix, then at least one of its eigenvalues must be zero. If A is positive (negative) definite, then A^{-1} is positive (negative) definite.*

Theorem A.2.7 *If both A and B are positive (negative) definite, and if $A - B$ is also positive (negative) definite, then $B^{-1} - A^{-1}$ is positive (negative) definite.*

Quadratic Forms with Nonsymmetric Matrices

Quadratic forms generally involve symmetric matrices. However, it is clear that equation (A.61) is well defined even when A is not symmetric. The form

$$x^T Ax = 0 \; \forall \, x \in \mathbb{R}^n$$

for some $A \in \mathbb{R}^{n \times n}$ occasionally occurs in derivations and deserves some attention. Before continuing, we note the following.

Theorem A.2.8 *If $A \in \mathbb{R}^{n \times n}$ is symmetric and*

$$x^T Ax = 0 \; \forall x \in \mathbb{R}^n,$$

then $A = 0$.

Proof:

$$x^T A x = \sum_{i=1}^{n} \sum_{j=1}^{n} a_{ij} x_i x_j = \sum_{i=1}^{n} a_{ii} x_i^2 + 2 \sum_{i=2}^{n} \sum_{j=1}^{i-1} a_{ij} x_i x_j = 0.$$

For this to be true for arbitrary x, all coefficients a_{ij} must be zero. ∎

Definition A.2.2 *The real matrix A is skew-symmetric if $A = -A^T$.*

For any skew-symmetric A, the diagonal elements are zero, since $a_{ii} = -a_{ii}$ only for $a_{ii} = 0$.

Theorem A.2.9 *For A skew-symmtric and any vector x of appropriate dimension,*

$$x^T A x = 0. \tag{A.78}$$

Proof:

$$x^T A x = x^T A^T x,$$

but by definition $A^T = -A$, so

$$x^T A x = -x^T A x,$$

and this can be true only if $x^T A x = 0$. ∎

The proof of the following is trivial.

Theorem A.2.10 *Any square matrix A can be written uniquely as the sum of a symmetric part A_s and a skew-symmetric part A_w.*

Given the above, we have

$$x^T A x = x^T (A_s + A_w) x = x^T A_s x$$

and

$$A + A^T = A_s + A_w + A_s^T + A_w^T = 2A_s.$$

As a result of these statements and Theorem A.2.8, we note the following.

Theorem A.2.11

$$x^T A x = 0 \; \forall x \in \mathbb{R}^n \implies A + A^T = 0. \tag{A.79}$$

It is not true, however, that $x^T A x = 0 \; \forall x \in \mathbb{R}^n \implies A = 0$, as the matrix may have a nonzero skew-symmetric part.

A.2.6 Time-Varying Vectors and Matrices

A time-varying column vector $x(t)$ is defined as a column vector whose components are themselves functions of time, i.e.,

$$x(t) = \begin{bmatrix} x_1(t) \\ x_2(t) \\ \vdots \\ x_n(t) \end{bmatrix}, \tag{A.80}$$

while a time-varying matrix $A(t)$ is defined as a matrix whose elements are time functions, i.e.,

$$A(t) = \begin{bmatrix} a_{11}(t) & a_{12}(t) & \cdots & a_{1m}(t) \\ a_{21}(t) & a_{22}(t) & \cdots & a_{2m}(t) \\ \vdots & \vdots & \vdots & \vdots \\ a_{n1}(t) & a_{n2}(t) & \cdots & a_{nm}(t) \end{bmatrix}. \tag{A.81}$$

The addition of time-varying vectors and matrices, their multiplication, and the scalar-product operations are defined as before.

Time Derivatives

The time derivative of the vector $x(t)$ is denoted by $d/dt\ x(t)$ or $\dot{x}(t)$ and is defined by

$$\frac{d}{dt}x(t) \triangleq \dot{x}(t) \triangleq \begin{bmatrix} \dot{x}_1(t) \\ \dot{x}_2(t) \\ \vdots \\ \dot{x}_n(t) \end{bmatrix}. \tag{A.82}$$

The time derivative of the matrix $A(t)$ is denoted by $d/dt \quad \mathbf{A}(t)$ or $\dot{A}(t)$ and is defined by

$$\frac{d}{dt}A(t) \triangleq \dot{A}(t) \triangleq \begin{bmatrix} \dot{a}_{11}(t) & \dot{a}_{12}(t) & \cdots & \dot{a}_{1m}(t) \\ \dot{a}_{21}(t) & \dot{a}_{22}(t) & \cdots & \dot{a}_{2m}(t) \\ \vdots & \vdots & \vdots & \vdots \\ \dot{a}_{n1}(t) & \dot{a}_{n2}(t) & \cdots & \dot{a}_{nm}(t) \end{bmatrix}. \tag{A.83}$$

Of course, in order for $\dot{x}(t)$ or $\dot{A}(t)$ to make sense, the derivatives $\dot{x}_i(t)$ and $\dot{a}_{ij}(t)$ must exist.

Integration

We can define the integrals of vectors and matrices in a similar manner. Thus,

$$\int_{t_0}^{t_f} x(t)dt \triangleq \begin{bmatrix} \int_{t_0}^{t_f} x_1(t)dt \\ \int_{t_0}^{t_f} x_2(t)dt \\ \vdots \\ \int_{t_0}^{t_f} x_n(t)dt \end{bmatrix}, \tag{A.84}$$

$$\int_{t_0}^{t_f} A(t)dt \triangleq \begin{bmatrix} \int_{t_0}^{t_f} A_{11}(t)dt & \cdots & \int_{t_0}^{t_f} A_{1m}(t)dt \\ \vdots & \vdots & \vdots \\ \int_{t_0}^{t_f} A_{n1}(t)dt & \cdots & \int_{t_0}^{t_f} A_{nm}(t)dt \end{bmatrix}. \tag{A.85}$$

A.2.7 Gradient Vectors and Jacobian Matrices

Let us suppose that $x_1, x_2, \ldots, x_n$ are real scalars which are the components of the column n-vector x:

$$x = \begin{bmatrix} x_1 \\ x_2 \\ \vdots \\ x_n \end{bmatrix}. \tag{A.86}$$

Now consider a scalar-valued function of the x_i,

$$f(x_1, x_2, \ldots, x_n) \triangleq f(x). \tag{A.87}$$

Clearly, f is a function mapping n-dimensional vectors to scalars:

$$f : \mathbb{R}^n \to \mathbb{R}. \tag{A.88}$$

Definition A.2.3 *The gradient of f with respect to the column n-vector x is denoted $\partial f(x)/\partial x$ and is defined by*

$$\frac{\partial f}{\partial x} \triangleq \frac{\partial}{\partial x} f(\mathbf{x}) \triangleq \left[\begin{array}{cccc} \frac{\partial f}{\partial x_1} & \frac{\partial f}{\partial x_2} & \cdots & \frac{\partial f}{\partial x_n} \end{array} \right], \tag{A.89}$$

so that the gradient is a row n-vector.

Note A.2.1 *We will also use the notation*

$$f_x \triangleq \frac{\partial f}{\partial x}$$

to denote the partial derivative.

Example A.2.2 *Suppose $f : \mathbb{R}^3 \to \mathbb{R}$ and is defined by*

$$f(x) = f(x_1, x_2, x_3) = x_1^2 x_2 e^{-x_3}; \tag{A.90}$$

then

$$\frac{\partial f}{\partial x} = \left[\begin{array}{ccc} 2x_1 x_2 e^{-x_3} & x_1^2 e^{-x_3} & -x_1 x_2 e^{-x_3} \end{array} \right]. \tag{A.91}$$

Again let us suppose that $x \in \mathbb{R}^n$. Let us consider a function g,

$$g : \mathbb{R}^n \to \mathbb{R}^m, \tag{A.92}$$

such that

$$\begin{aligned} y = g(x), \quad x &\in \mathbb{R}^n, \\ y &\in \mathbb{R}^m. \end{aligned} \tag{A.93}$$

By this we mean

$$y_1 = g_1(x_1, x_2, \ldots, x_n) = g_1(x), \tag{A.94}$$

$$y_2 = g_2(x_1, x_2, \ldots, x_n) = g_2(x), \tag{A.95}$$

$$\vdots \tag{A.96}$$

$$y_m = g_m(x_1, x_2, \ldots, x_n) = g_m(x). \tag{A.97}$$

Definition A.2.4 *The Jacobian matrix of g with respect to x is denoted by $\partial g(x)/\partial x$ and is defined as*

$$\frac{\partial g(x)}{\partial x} \triangleq \begin{bmatrix} \frac{\partial g_1}{\partial x_1} & \frac{\partial g_1}{\partial x_2} & \cdots & \frac{\partial g_1}{\partial x_n} \\ \frac{\partial g_2}{\partial x_1} & \frac{\partial g_2}{\partial x_2} & \cdots & \frac{\partial g_2}{\partial x_n} \\ \vdots & \vdots & \vdots & \vdots \\ \frac{\partial g_m}{\partial x_1} & \frac{\partial g_m}{\partial x_2} & \cdots & \frac{\partial g_m}{\partial x_n} \end{bmatrix}. \tag{A.98}$$

Thus, if $g : \mathbb{R}^n \to \mathbb{R}^m$, its Jacobian matrix is an $m \times n$ matrix.

As an immediate consequence of the definition of a gradient vector, we have

$$\frac{\partial}{\partial x}\langle x, y\rangle = \frac{\partial y^T x}{\partial x} = y^T, \tag{A.99}$$

$$\frac{\partial}{\partial x}\langle x, Ay\rangle = (Ay)^T, \tag{A.100}$$

$$\frac{\partial}{\partial x}\langle Ax, y\rangle = \frac{\partial}{\partial x}\langle x, A^T y\rangle = y^T A. \tag{A.101}$$

The definition of a Jacobian matrix yields the relation

$$\frac{\partial}{\partial x} Ax = A. \tag{A.102}$$

Now suppose that $x(t)$ is a time-varying vector and that $f(x)$ is a scalar-valued function of x. Then by the chain rule

$$\frac{d}{dt} f(x) = \frac{\partial f}{\partial x_1}\dot{x}_1 + \frac{\partial f}{\partial x_2}\dot{x}_2 + \cdots + \frac{\partial f}{\partial x_n}\dot{x}_n = \sum_{i=1}^{n} \frac{\partial f}{\partial x_i}\dot{x}_i, \tag{A.103}$$

which yields

$$\frac{d}{dt}f(x) = \left\langle \left(\frac{\partial f}{\partial x}\right)^T, \dot{x}(t) \right\rangle. \tag{A.104}$$

Similarly, if $g : \mathbb{R}^n \to \mathbb{R}^m$, and if $x(t)$ is a time-varying column vector, then

$$\frac{d}{dt}g(x) \triangleq \begin{bmatrix} \frac{d}{dt}g_1(x) \\ \frac{d}{dt}g_2(x) \\ \vdots \\ \frac{d}{dt}g_m(x) \end{bmatrix} = \begin{bmatrix} \left\langle \left(\frac{\partial g_1}{\partial x}\right)^T, \dot{x}(t) \right\rangle \\ \left\langle \left(\frac{\partial g_2}{\partial x}\right)^T, \dot{x}(t) \right\rangle \\ \vdots \\ \left\langle \left(\frac{\partial g_m}{\partial x}\right)^T, \dot{x}(t) \right\rangle \end{bmatrix} = \left(\frac{\partial g}{\partial x}\right)\dot{x}(t). \tag{A.105}$$

It should be clear that gradient vectors and matrices can be used to compute mixed time and partial derivatives.

A.2.8 Second Partials and the Hessian

Consider once more a scalar function of a vector argument $f : \mathbb{R}^n \to \mathbb{R}$. The *Hessian* of f is the matrix of second partial derivatives of f with respect to the elements of x:

$$f_{xx} = \begin{bmatrix} \frac{\partial^2 f}{\partial x_1^2} & \frac{\partial^2 f}{\partial x_1 \partial x_2} & \cdots & \frac{\partial^2 f}{\partial x_1 \partial x_n} \\ \frac{\partial^2 f}{\partial x_2 \partial x_1} & \frac{\partial^2 f}{\partial x_2^2} & \cdots & \frac{\partial^2 f}{\partial x_2 \partial x_n} \\ \vdots & \vdots & \ddots & \vdots \\ \frac{\partial^2 f}{\partial x_n \partial x_1} & \frac{\partial^2 f}{\partial x_n \partial x_2} & \cdots & \frac{\partial^2 f}{\partial x_n^2} \end{bmatrix}. \tag{A.106}$$

It is clear from the definition that the Hessian is symmetric.

Consider the function $f(x, u) : \mathbb{R}^{n+m} \to \mathbb{R}$. As above, the partial of the function with respect to one of the vector arguments is a row vector, as in

$$f_x = \frac{\partial f}{\partial x} = \begin{bmatrix} \frac{\partial f}{\partial x_1} & \frac{\partial f}{\partial x_2} & \cdots & \frac{\partial f}{\partial x_n} \end{bmatrix}.$$

The matrix of second partials of this with respect to the vector u is the $n \times m$ matrix given by

$$f_{xu} \triangleq \frac{\partial}{\partial u} f_x^T = \begin{bmatrix} \frac{\partial^2 f}{\partial x_1 \partial u_1} & \frac{\partial^2 f}{\partial x_1 \partial u_2} & \cdots & \frac{\partial^2 f}{\partial x_1 \partial u_m} \\ \frac{\partial^2 f}{\partial x_2 \partial u_1} & \frac{\partial^2 f}{\partial x_2 \partial u_2} & \cdots & \frac{\partial^2 f}{\partial x_2 \partial u_m} \\ \vdots & \vdots & \ddots & \vdots \\ \frac{\partial^2 f}{\partial x_n \partial u_1} & \frac{\partial^2 f}{\partial x_n \partial u_2} & \cdots & \frac{\partial^2 f}{\partial x_n \partial u_m} \end{bmatrix}. \tag{A.107}$$

A.2.9 Vector and Matrix Norms

We conclude our brief introduction to column vectors, matrices, and their operations by discussing the concept of the norm of a column vector and the norm of a matrix. The norm is a generalization of the familiar magnitude of Euclidean length of a vector. Thus, the norm is used to decide how large a vector is and also how large a matrix is; in this manner it is used to attach a scalar magnitude to such multivariable quantities as vectors and matrices.

Norms for Column Vectors

Let us consider a column n-vector x;

$$x = \begin{bmatrix} x_1 \\ x_2 \\ \vdots \\ x_n \end{bmatrix}. \tag{A.108}$$

The Euclidean norm of x, denoted by $||x||_2$, is simply defined by

$$||x||_2 = (x_1^2 + x_2^2 + \cdots + x_n^2)^{1/2} = \sqrt{\langle x, x \rangle}. \tag{A.109}$$

It should be clear that the value of $||x||_2$ provides us with an idea of how big x is. We recall that the Euclidean norm of a column n-vector satisfies the following conditions:

$$||x||_2 \;\geq\; 0 \quad \text{and} \quad ||x||_2 = 0 \quad \text{if and only if} \quad x = 0, \tag{A.110}$$

$$||\alpha x||_2 = |\alpha| \cdot ||x||_2 \quad \text{for all scalars } \alpha, \tag{A.111}$$

$$||x+y||_2 \leq ||x||_2 + ||y||_2, \quad \text{the triangle inequality.} \tag{A.112}$$

For many applications, the Euclidean norm is not the most convenient to use in algebraic manipulations, although it has the most natural geometric interpretation. For this reason, one can generalize the notion of a norm in the following way.

Definition A.2.5 *Let x and y be column n-vectors. Then a scalar-valued function of x qualifies as a norm $||x||$ of x provided that the following three properties hold:*

$$||x|| > 0 \qquad \forall\, x \neq 0, \tag{A.113}$$

$$||\alpha x|| = |\alpha| \cdot ||x|| \qquad \forall\, \alpha \in \mathbb{R}, \tag{A.114}$$

$$||x+y|| \leq ||x|| + ||y|| \qquad \forall\, x, y. \tag{A.115}$$

The reader should note that Equations (A.113) to (A.115) represent a consistent generalization of the properties of the Euclidean norm given in Equations (A.110) to (A.112).

In addition to the Euclidean norm, there are two other common norms:

$$||x||_1 = \sum_{i=1}^{n} |x_i|, \tag{A.116}$$

$$||x||_\infty = \max_i |x_i|. \tag{A.117}$$

We encourage the reader to verify that the norms defined by (A.116) and (A.117) indeed satisfy the properties given in Equations (A.110) to (A.112).

Example A.2.3 *Suppose that x is the column vector*

$$x = \begin{bmatrix} 2 \\ -1 \\ 3 \end{bmatrix}; \tag{A.118}$$

then $||x||_1 = |2| + |-1| + |3| = 6$, $||x||_2 = (4+1+9)^{1/2} = \sqrt{14}$, $||x||_\infty = \max\{|2|, |-1|, |3|\} = 3$.

Matrix Norms

Next we turn our attention to the concept of a norm of a matrix. To motivate the definition we simply note that a column n-vector can also be viewed as an $n \times 1$ matrix. Thus, if we are to extend the properties of vector norms to those of the matrix norms, they should be consistent. For this reason, we have the following definition.

Definition A.2.6 *Let A and B be real $n \times m$ matrices with elements a_{ij} and b_{ij} $(i = 1, 2, \ldots, n$: $j = 1, 2, \ldots, m)$. Then the scalar-valued function $||A||$ of A qualifies as the norm of A if the following properties hold:*

$$||A|| > 0 \quad \text{provided not all} \quad a_{ij} = 0, \tag{A.119}$$

$$||\alpha A|| = |\alpha| \cdot ||A|| \quad \forall\, \alpha \in \mathbb{R}, \tag{A.120}$$

$$||A + B|| \leq ||A|| + ||B||. \tag{A.121}$$

As with vector norms, there are many convenient matrix norms, e.g.,

$$||A||_1 = \sum_{i=1}^{n} \sum_{j=1}^{m} |a_{ij}|, \tag{A.122}$$

$$||A||_2 = \left(\sum_{i=1}^{n} \sum_{j=1}^{m} a_{ij}^2 \right)^{1/2}, \tag{A.123}$$

$$||A||_\infty = \max_i \sum_{j=1}^{m} |a_{ij}|. \tag{A.124}$$

Once more we encourage the reader to prove that these matrix norms do indeed satisfy the defining properties of Equations (A.119) to (A.121).

Properties

Two important properties that hold between norms which involve multiplication of a matrix with a vector and multiplication of two matrices are summarized in the following two theorems.

Theorem A.2.12 *Let A be an $n \times m$ matrix with real elements a_{ij} $(i = 1, 2, \ldots, n$: $j = 1, 2, \ldots, m)$. Let x be a column m-vector with elements x_j $(j = 1, 2, \ldots, m)$. Then*

$$||Ax|| \leq ||A|| \cdot ||x|| \tag{A.125}$$

in the sense that

$$\text{(a)} \qquad ||Ax||_1 \leq ||A||_1 \cdot ||x||_1, \tag{A.126}$$

$$\text{(b)} \qquad ||Ax||_2 \leq ||A||_2 \cdot ||x||_2, \tag{A.127}$$

$$\text{(c)} \qquad ||Ax||_\infty \leq ||A||_\infty \cdot ||x||_\infty. \tag{A.128}$$

Proof: Let $y = Ax$; then y is a column vector with n-components $y_1, y_2, \ldots, y_n$.

$$\begin{aligned}
\text{(a)} \qquad ||Ax||_1 &= ||y||_1 \triangleq \sum_{i=1}^{n} |y_i| = \sum_{i=1}^{n} \left| \sum_{j=1}^{m} a_{ij} x_j \right| \\
&\leq \sum_{i=1}^{n} \sum_{j=1}^{m} |a_{ij} x_j| = \sum_{i=1}^{n} \sum_{j=1}^{m} |a_{ij}||x_j| \\
&\leq \sum_{i=1}^{n} \sum_{j=1}^{m} |a_{ij}| \cdot ||x||_1 \quad \text{since} \quad ||x||_1 \geq |x_j| \\
&= \left(\sum_{i=1}^{n} \sum_{j=1}^{m} |a_{ij}| \right) ||x||_1 = ||A||_1 \cdot ||x||_1.
\end{aligned}$$

$$\begin{aligned}
\text{(b)} \qquad ||Ax||_2 &= ||y||_2 \triangleq \left(\sum_{i=1}^{n} |y_i| \right)^{1/2} = \left[\sum_{i=1}^{n} \left(\sum_{j=1}^{m} a_{ij} x_j \right)^2 \right]^{1/2} \\
&\leq \left[\sum_{i=1}^{n} \left(\sum_{j=1}^{m} a_{ij}^2 \right) \left(\sum_{j=1}^{m} x_j^2 \right) \right]^{1/2} \quad \text{by the Schwartz inequality} \\
&= \left[\sum_{i=1}^{n} \left(\sum_{j=1}^{m} a_{ij}^2 \right) ||x||_2^2 \right]^{1/2} = \left(\sum_{i=1}^{n} \sum_{j=1}^{m} a_{ij}^2 \right)^{1/2} ||x||_2 \\
&= ||A||_2 \cdot ||x||_2.
\end{aligned}$$

$$
\begin{aligned}
\text{(c)} \quad ||AX||_\infty &= ||y||_\infty \triangleq \max_i |y_i| = \max_i \left| \sum_{j=1}^{m} a_{ij}x_j \right| \\
&\le \max_i \left(\sum_{j=1}^{m} |a_{ij}x_j| \right) = \max_i \left(\sum_{j=1}^{m} |a_{ij}||x_j| \right) \\
&\le \max_i \left(\sum_{j=1}^{m} |a_{ij}| \cdot ||x||_\infty \right), \qquad \text{because } ||x||_\infty \ge |x_j|, \\
&= \max_i \left(\sum_{j=1}^{m} |a_{ij}| \right) ||x||_\infty = ||A||_\infty ||x||_\infty.
\end{aligned}
$$

∎

We shall leave it to the reader to verify the following theorem by imitating the proofs of Theorem A.2.12.

Theorem A.2.13 *Let A be a real $n \times m$ matrix and let B be a real $m \times q$ matrix; then*

$$||AB|| \le ||A|| \cdot ||B|| \tag{A.129}$$

in the sense that

$$\text{(a)} \quad ||AB||_1 \le ||A||_1 \cdot ||B||_1, \tag{A.130}$$

$$\text{(b)} \quad ||AB||_2 \le ||A||_2 \cdot ||B||_2, \tag{A.131}$$

$$\text{(c)} \quad ||AB||_\infty \le ||A||_\infty \cdot ||B||_\infty. \tag{A.132}$$

A multitude of additional results concerning the properties of norms are available.

Spectral Norm

A very useful norm, called the *spectral norm*, of a matrix is denoted by $||A||_s$. Let A be a real $n \times m$ matrix. Then A^T is an $m \times n$ matrix, and the product matrix A^TA is an $m \times m$ real matrix. Let us compute the eigenvalues of A^TA, denoted by $\lambda_i(A^TA)$, $i = 1, 2, \ldots, m$. Since the matrix A^TA is symmetric and positive semidefinite, it has real nonnegative eigenvalues, i.e.,

$$\lambda_i(A^T A) \geq 0, \qquad i = 1, 2, \ldots, m. \tag{A.133}$$

Then the spectral norm of A is defined by

$$||A||_s = \max[\lambda_i(A^T A)]^{1/2}, \tag{A.134}$$

i.e., it is the square root of the maximum eigenvalue of $A^T A$.

Remark A.2.1 *The singular values of A are given by $\lambda^{1/2}(A^T A)$.*

A.2.10 Taylor's Theorem for Functions of Vector Arguments

We consider the Taylor expansion of a function of a vector argument. Using the definitions developed above, we have the following.

Theorem A.2.14 *Let $f(x) : \mathbb{R}^n \mapsto \mathbb{R}$ be N times continuously differentiable in all of its arguments at a point x_0. Consider $f(x_0+\varepsilon h)$, where $||h|| = 1$. (It is not important which specific norm is used; for simplicity, we may assume it is the Euclidean.) Then*

$$f(x_0 + \varepsilon h) = f(x_0) + \varepsilon \left.\frac{\partial f}{\partial x}\right|_{x_0} h + \frac{\varepsilon^2}{2!} h^T \left.\frac{\partial^2 f}{\partial x^2}\right|_{x_0} h + \cdots + R_N,$$

where

$$\lim_{\varepsilon \to 0} \frac{R_N}{\varepsilon^N} \to 0.$$

The coefficients of the first two terms in the expansion are the gradient vector and the Hessian matrix.

A.3 Linear Dynamical Systems

In this section we review briefly some results in linear systems theory that will be used in the main text. For a more complete treatment, see [8] or [12]. Consider the continuous-time linear system

$$\dot{x}(t) = A(t)x(t) + B(t)u(t), \qquad x(t_0) = x_0 \text{ given}, \tag{A.135}$$

where $x(\cdot) \in \mathbb{R}^n, u(\cdot) \in \mathbb{R}^m$, and A and B are appropriately dimensioned real matrices. The functions $a_{ij}(t), i = 1, \ldots, n, j = 1, \ldots, n$, that make up $A(t)$ are continuous, as are the elements of $B(t)$. The control functions in $u(\cdot)$ will be restricted to being piecewise continuous and everywhere defined.

In most cases, for both clarity and convenience, we will drop the explicit dependence of the variables on time. When it is important, it will be included.

Some Terminology. The vector x will be termed the *state vector* or simply the *state*, and u is the *control* vector. The matrix A is usually known as the *plant* or the *system* matrix, and B is the *control coefficient* matrix.

We will assume that the system is always *controllable*. This means that given $x_0 = 0$ and some desired final state x_1 at some final time $t_1 > t_0$, there exists some control function $u(t)$ on the interval $[t_0, t_1]$ such that $x(t_1) = x_1$ using that control input. This is discussed in detail in [8].

Fact: Under the given assumptions, the solution $x(\cdot)$ associated with a particular x_0 and control input $u(\cdot)$ is unique. In particular, for $u(\cdot) \equiv 0$ and any specified t_1 and x_1, there is **exactly one initial condition $\boldsymbol{x_0}$ and one associated solution of (A.135) such that $\boldsymbol{x(t_1) = x_1}$.**

State Transition Matrix. Consider the system with the control input identically zero. Then we have

$$\dot{x}(t) = A(t)x(t).$$

Under this condition, we can show that the solution $x(t)$ is given by the relation

$$x(t) = \Phi(t, t_0)x_0,$$

where $\Phi(\cdot, \cdot)$ is known as the *state transition matrix*, or simply the *transition matrix* of the system. The state transition matrix obeys the differential equation

$$\frac{d}{dt}\Phi(t, t_0) = A(t)\Phi(t, t_0), \qquad \Phi(t_0, t_0) = I, \tag{A.136}$$

and has a couple of obvious properties:

1. $\Phi(t_2, t_0) = \Phi(t_2, t_1)\Phi(t_1, t_0)$,
2. $\Phi(t_2, t_1) = \Phi^{-1}(t_1, t_2)$.

It is important to note that the transition matrix is independent of the initial state of the system.

In the special case of $A(t) = A$ constant, the state transition matrix is given by

$$\Phi(t, t_0) = e^{A(t-t_0)}, \tag{A.137}$$

where the matrix exponential is defined by the series

$$e^{A(t-t_0)} = I + A(t - t_0) + \frac{1}{2!}A^2(t - t_0)^2 + \cdots + \frac{1}{k!}A^k(t - t_0)^k + \cdots.$$

It is easy to see that the matrix exponential satisfies (A.136).

Fundamental Matrix: A *fundamental matrix* of the system (A.135) is any matrix $X(t)$ such that

$$\frac{d}{dt}X(t) = A(t)X(t). \tag{A.138}$$

The state transition matrix is the fundamental matrix that satisfies the initial condition $X(t_0) = I$.

Fact: If there is any time t_1 such that a fundamental matrix is nonsingular, then it is nonsingular for all t. An obvious corollary is that the state transition matrix is nonsingular for all t (since the initial condition I is nonsingular).

Fact: If $X(t)$ is any fundamental matrix of $\dot{x} = Ax$, then for all $t, t_1 \in \mathbb{R}$,

$$\Phi(t, t_1) = X(t) \cdot X^{-1}(t_1).$$

Solution to the Linear System. The solution to the linear system for some given control function $u(\cdot)$ is given by

$$x(t) = \Phi(t, t_0)x_0 + \int_{t_0}^{t} \Phi(t, \tau)B(\tau)u(\tau)d\tau. \tag{A.139}$$

The result is proven by taking the derivative and showing that it agrees with (A.135).

Bibliography

[1] Anderson, B. D. O., and Moore, J. B. *Linear Optimal Control.* Prentice Hall, Englewood Cliffs, NJ, 1971.

[2] Athans, M., Ed., *Special Issue on the Linear-Quadratic-Gaussian Problem.* IEEE Transactions on Automatic Control, Vol. AC-16, December 1971.

[3] Athans, M., Dertouzos, M. L., Spann, R. N., and Mason, S. J. *Systems, Networks, and Computation.* McGraw-Hill, New York, 1974.

[4] Bell, D. J., and Jacobson, D. H., *Singular Optimal Control Problem.* Academic Press, New York, 1975.

[5] Betts, J., *Practical Methods for Optimal Control Using Nonlinear Programming.* SIAM, Philadelphia, 2001.

[6] Bliss, G. A., *Lectures on the Calculus of Variations.* University of Chicago Press, Chicago, 1946.

[7] Breakwell, J. V., Speyer, J. L., and Bryson, A. E. *Optimization and Control of Nonlinear Systems Using the Second Variation.* SIAM Journal on Control, Series A, Vol. 1, No. 2, 1963, pp. 193–223.

[8] Brockett, R. W., *Finite Dimensional Linear Systems.* John Wiley, New York, 1970.

[9] Broyden, C. G. *The Convergence of a Class of Double-Rank Minimization Algorithms.* Journal of the Institute of Mathematics and Its Applications, Vol. 6, 1970, pp. 76–90.

[10] Bryant, G. F. and Mayne, D. Q. *The Maximum Principle.* International Journal of Control, Vol. 20, No. 6, 1174, pp. 1021–1054.

[11] Bryson, A. E., and Ho, Y. C. *Applied Optimal Control.* Hemisphere Publishing, Washington, D.C., 1975.

[12] Callier, F. M., and Desoer C. A., *Linear Systems Theory*, Springer Texts in Electrical Engineering, Springer, New York, 1994.

[13] Cannon, M. D., Cullum, C. D., and Polak, E., *Theory of Optimal Control and Mathematical Programming.* McGraw-Hill, New York, 1970.

[14] Clements, D. J., and Anderson, B. D. O., *Singular Optimal Control: The Linear Quadratic Problem.* Springer, New York, 1978.

[15] Coddington, E., and Levinson, N., *Theory of Ordinary Differential Equations.* McGraw–Hill, New York, 1958.

[16] Cullen, C. G., *Matrices and Linear Transformations*, 2nd Edition, Dover, New York, 1990.

[17] Davidon, W. C. *Variable Metric Method for Minimization.* SIAM Journal on Optimization, Vol. 1, 1991, pp. 1–17.

[18] Doyle, J. C., Glover, K., Khargonekar, P. P. and Francis, B., *State-Space Solutions to Standard H-2 and H-Infinity Control-Problems*, IEEE Transactions on Automatic Control, Vol. 34, No. 8, August 1989, pp. 831–847.

[19] Dreyfus, S. E., *Dynamic Programming and the Calculus of Variations.* Academic Press, New York, 1965.

[20] Dyer, P., and McReynolds, S. R., *The Computation and Theory of Optimal Control.* Academic Press, New York, 1970.

[21] Friedman, A., *Advanced Calculus.* Holt, Rinehart and Winston, New York, 1971.

[22] Gelfand, I. M., and Fomin, S. V., *Calculus of Variations.* Prentice Hall, Englewood Cliffs, NY, 1963.

[23] Gill, P. E., Murray, W., and Wright, M. H., *Practical Optimization,* Academic Press, New York, 1981.

[24] Grtschel, M., Krumke, S. O. and Rambau, J., Eds., *Online Optimization of Large Scale Systems.* Springer-Verlag, Berlin, 2001.

[25] Halkin, H., *Mathematical Foundations of System Optimization.* Topics in 0ptimization, G. Leitmann, Ed., Academic Press, New York, 1967.

[26] Hestenes, M. R., *Calculus of Variations and Optimal Control Theory.* John Wiley, New York, 1965.

[27] Hohn, F. E., *Elementary Matrix Algebra.* 3rd Edition, Macmillan, New York, 1973.

[28] Jacobson, D. H., *A Tutorial Introduction to Optimality Conditions in Nonlinear Programming.* 4th National Conference of the Operations Research Society of South Africa, November 1972.

[29] Jacobson, D. H. *Extensions of Linear-Quadratic Control, Optimization and Matrix Theory,* Academic Press, New York, 1977.

[30] Jacobson, D. H., Lele, M. M., and Speyer, J. L., *New Necessary Conditions of Optimality for Control Problems with State Variable Inequality Constraints,* Journal of Mathematical Analysis and Applications., Vol. 35, No. 2, 1971, pp. 255–284 .

[31] Jacobson, D. H., Martin, D. H., Pacher, M., and Geveci, T. *Extensions of Linear-Quadratic Control Theory,* Lecture Notes in Control and Information Sciences, Vol. 27, Springer-Verlag, Berlin, 1980.

[32] Jacobson, D. H., and Mayne, D. Q., *Differential Dynamic Programming.* Elsevier, New York, 1970.

[33] Kuhn, H., and Tucker, A. W. *Nonlinear Programming.* Second Berkeley Symposium of Mathematical Statistics and Probability, University of California Press, Berkeley, 1951.

[34] Kwakernaak, H., and Sivan, R. *Linear Optimal Control Systems.* Wiley Interscience, New York, 1972.

[35] Mangasarian, O. L., *Nonlinear Programming.* McGraw-Hill, New York, 1969.

[36] Nocedal, J., and Wright, S. J., *Numerical Optimization,* Springer Series in Operations Research, P. Glynn and S. M. Robinson, Eds., Springer-Verlag, New York, 2000.

[37] Pars, L., *A Treatise on Analytical Dynamics.* John Wiley, New York, 1965.

[38] Pontryagin, L. S., Boltyanskii, V. G., Gamkrelidze, R. V., and Mischenko, E. F., *The Mathematical Theory of Optimal Processes,* L.W. Neustadt, Ed., Wiley Interscience, New York, 1962.

[39] Rall, L. B., *Computational Solution of Nonlinear Operator Equations.* Wiley, New York, 1969.

[40] Rhee, I., and Speyer, J. L., *A Game-Theoretic Approach to a Finite-Time Disturbance Attenuation Problem*, IEEE Transactions on Automatic Control, Vol. AC-36, No. 9, September 1991, pp. 1021–1032.

[41] Rodriguez-Canabal, J., *The Geometry of the Riccati Equation.* Stochastics, Vol. 1, 1975, pp. 347–351.

[42] Sain, M., Ed., *Special Issue on Multivariable Control.* IEEE Transactions on Automatic Control, Vol. AC-26, No. 1, February 1981.

[43] Stoer, J., and Bulirsch, R., *Introduction to Numerical Analysis*, 3rd Edition, Texts in Applied Mathematics 12, Springer, New York, 2002.

[44] Varaiya, P. P., *Notes on Optimization.* Van Nostrand, New York, 1972.

[45] Willems, J. D., *Least Squares Stationary Optimal Control and the Algebraic Riccati Equation.* IEEE Transactions on Automatic Control, Vol. AC-16, December 1971, pp. 621–634.

[46] Wintner, A., *The Analytical Foundations of Celestial Mechanics.* Princeton University Press, Princeton, NJ, 1947.

[47] Zangwill, W., *Nonlinear Programming: A Unified Approach.* Prentice–Hall, Englewood Cliffs, NJ, 1969.

Index